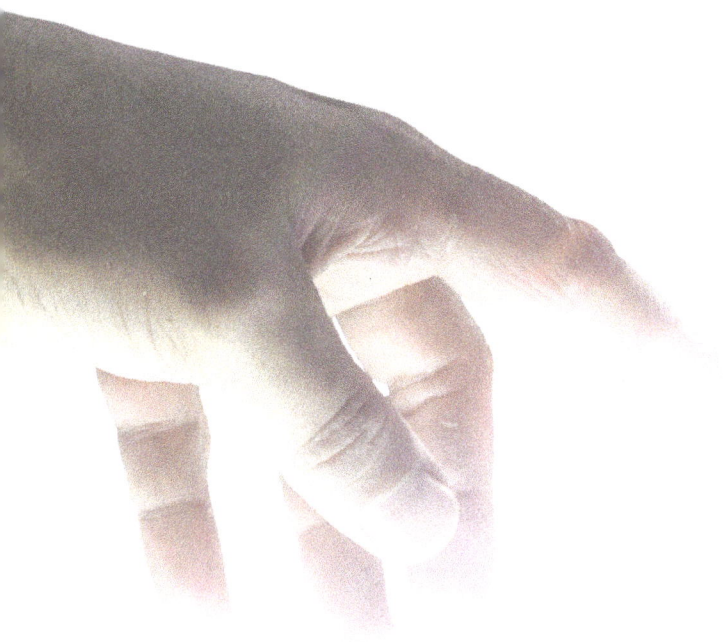

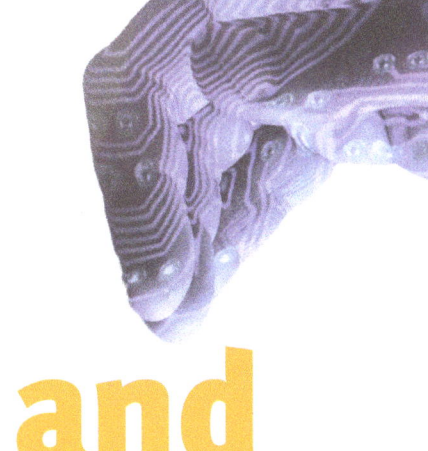

SCIENCE and CHRISTIANITY

Understanding the conflict myth

CHRIS MULHERIN

This book is dedicated to the memory of Ben Mulherin, 24/8/1985 – 8/12/2008.
– Chris Mulherin

Published in Australia by
Garratt Publishing
32 Glenvale Crescent
Mulgrave, VIC 3170
www.garrattpublishing.com.au

All rights reserved. Except as provided by the Australian copyright law, no part of this book may be reproduced in any way without permission in writing from the publisher.

Text copyright © Chris Mulherin 2019
Series Design by Lynne Muir
Design and typesetting by Guy Holt
Images copyright © iStock

Attributions List for images from Wikimedia Creative Commons:
William Whewell, page 13
Ptolemaic system, page 13
Dhatfi eld, 2008, Diagram of the Schrödinger's cat thought experiment, page 18
NASA/Paul E. Alers, 2008, CC BY-SA 3.0, Dr Stephen Hawking, page 26
Cstreet, 2005, Professor Richard Dawkins, page 26
NASA, ESA, Hubble Heritage Team (STScI/AURA), 2014, Pillars of Creation, page 29
Ken Crawford, 2011, Horsehead Nebula, page 30
Thilo Parg, CC BY-SA 3.0, Oetzi the Iceman Reconstruction, page 32
Noli Me Tangere by Antonio da Correggio, page 44
Jvangiel, 2013, CC BY-SA 3.0, Lawrence Krauss, page 54
Galileo Facing the Roman Inquisition by Cristiano Banti, page 56
Dialogue Concerning the Two Chief World Systems, page 57
Urbanus VIII by Pietro da Cortona, page 57
Jjstott, 2010, CC BY-SA 3.0, Crab Nebula, page 59
H. Raab (user: Vesta), 2018, CC BY-SA 4.0, Vatican Observatory, page 59

Scripture quotations are drawn from the New Revised Standard Version of the Bible, copyright © 1989 by the Division of Christian Education of the National Council of the Churches of Christ in the USA. Used by permission. All rights reserved.

ISBN 9781925073515

Cataloguing in Publication information for this title is available from the
National Library of Australia.
www.nla.gov.au

The author and publisher gratefully acknowledge the permission granted to reproduce the copyright material in this book. Every effort has been made to trace copyright holders and to obtain their permission for the use of copyright material.

The publisher apologises for any errors or omissions in the above list and would be grateful if notified of any corrections that should be incorporated in future reprints or editions of this book.

CONTENTS

Introduction — 1
- Scope of this book — 3
- Summary of this book — 4
- **Article 1** The Bible is not science: St Augustine's warning about confusing science and faith — 5
- **Article 2** Watch your genre! The importance of understanding text type — 6

CHAPTER ONE
The wonders and benefits of science — 9
- The surprises and benefits of science — 10
- The history of science — 12
- **Article 3** The weird, weird world of quantum physics — 18
- **Article 4** Visionary figures in science — 19
- **Article 5** Charles Darwin affirms God's two books — 21

CHAPTER TWO
The limits of science — 23
- Science has its limits — 24
- Practical limits of science — 24
- Philosophical limits of science — 25
- Can science offer certainty? — 27
- **Article 6** The Hubble Space Telescope: Looking back in time — 29
- **Article 7** Dr Jennifer Wiseman: Astrophysicist and public speaker — 30
- **Article 8** DNA, the instructions for life — 31
- **Article 9** Professor Phil Batterham: Past President of the International Genetics Federation — 32
- **Article 10** Climate science and Christian belief: A case study — 33

CHAPTER THREE
The problem of scientism — 37
- What's in a worldview? — 38
- Naturalism vs naturalism — 40
- The problems of scientism: Taking science beyond its limits — 42
- What about miracles? Did Jesus rise from the dead? — 43
- **Article 11** What is philosophy of science? — 45

CHAPTER FOUR
"I believe in science, so ..."
The myth of conflict — 47
- The sources and celebrities of the New Atheism — 48
- Rumours of divorce: Understanding the conflict thesis — 48
- **Article 12** A New Atheist example: Lawrence Krauss and *A Universe from Nothing* — 52
- **Article 13** Evolution and the Bible — 55
- **Article 14** Clearing the air about the Galileo affair — 56
- **Article 15** Dr Francis Collins: Medical researcher and genetic pioneer — 58
- **Article 16** The Roman Catholic Church and science — 59
- **Article 17** Sir John Polkinghorne: Theoretical physicist and Anglican priest — 60

Conclusion — 63

End notes and resources — 65

Introduction

Welcome to a journey of exploration into some of the most significant questions we can ask ourselves. Some of these are scientific questions about the wonders of the physical universe we find ourselves in. Other issues are philosophical and religious: they are concerned with the meaning and purpose of life on this dust-speck planet called Earth, which floats through an unimaginably vast cosmos.

INTRODUCTION

Have you ever gazed into the night sky at the stars? Have you ever asked yourself questions that make you feel small and insignificant in the vastness of space and time? For thousands of years, we human beings have explored the world around us and been fascinated by the awesome marvels of the universe. We love to ask about how the natural world works: What are stars? What are they made of? How many are there? Why do they move? And closer to home: Why do things fall down? What is the smallest thing there is? Is everything made of the same 'stuff'? What is light? How old is the universe? And how did it come to be?

Almost 3000 years ago a Hebrew songwriter started Psalm 19 this way:

> The heavens are telling the glory of God; and the firmament proclaims his handiwork. Day to day pours forth speech, and night to night declares knowledge (Psalm 19:1–2).

For the psalm-writer, something amazing about God is revealed in "his handiwork". For the psalmist, the universe 'speaks' of its creator. In another psalm, the psalmist ponders the majesty of the creation and wonders at God's care for human beings:

> O Lord, our Sovereign, how majestic is your name in all the earth! You have set your glory above the heavens … When I look at your heavens, the work of your fingers, the moon and the stars that you have established; what are human beings that you are mindful of them, mortals that you care for them? (Psalm 8:1, 3–4).

This theme of seeing the natural world as the work of God, revealing something about its creator, has an ancient history that goes back long before the time of Christ. And, in the Christian era, this understanding of 'nature' as the creation of God served as one of the inspirations for science as we know it today. Although it is very difficult to define exactly the boundaries of science, in this book science is taken to mean the systematic study of the natural, material world, using theory and experiments, such as we see in physics, chemistry, biology, geology, astronomy, and so on. In Chapter 1 we will think more about the nature of science.

For many of the great scientists of history, their Christian belief was a motivation for exploring the wonders of the universe – a universe often referred to as "the book of God's works". However, while the harmony of science and Christianity has been the normal view through Christian history, the possibility of conflict between the two has always been present – sometimes quietly and sometimes very loudly indeed. While Christianity and science have mostly been friends for a long, long time, the prospect of conflict has a long history, too. In fact, it is over 1500 years ago that one of Christianity's greatest thinkers, St Augustine of Hippo, warned Christians about getting into conflict with science. Although St Augustine did not use the word 'science', he encouraged believers not to think that science and Christianity were in conflict. (See Article #1, "The Bible is not science: St Augustine's warning about confusing science and faith" on page 5). We will return to the 'conflict thesis' in Chapter 4. First we will look at the nature of science, as well as the history of the relationship between science and Christianity.

This theme of seeing the natural world as the work of God, revealing something about its creator, has an ancient history that goes back long before the time of Christ.

Scope of this book

This book is a short introduction to science and Christian belief in the part of the world historically known as the Western world, or 'the West'. The West includes those countries and cultures dominated by European values and thinking. Of course, both science and Christianity have long histories in other cultures, but this book is written with a Western audience in mind. The West has inherited what is sometimes called the Judaeo-Christian tradition: Jesus was a Jew and his followers soon came to be called Christians. Both Jews and Christians (and Muslims, too, later in history) affirm 'ethical monotheism': there is one God and God sets the standard for living a moral life. In just the few centuries after Jesus, Christianity spread across Europe and north Africa and parts of Asia, and it continued to spread all around the world in the centuries that followed. Today, Christianity is the largest religious group on the planet, and includes just under one-third of the global population.

Notice that this book is not about science and 'religion'. It is notoriously difficult to define 'religion'; there are many religions, some believe in a god or many gods, and some do not have a god or gods. As we will see in Chapter 3, Christianity might be described as a worldview while many religions are not. For the sake of this book, it is important to clarify that the topic is about the relationship between science and Christian belief, expressed for example in the Nicene Creed.

Science as we know it today was nurtured, and it thrived, in this Western and mostly Christian context. This book is about two of the most powerful cultural forces in the Western world, and the relationship between them. However, while modern science was matured in a Christian context, it has roots in Greek philosophy, especially as preserved and amplified by Islamic scholars in the late first millennium.

Until recently (historically speaking), science in the West has walked hand-in-hand with Christianity, so this book will explore some of that long and fruitful relationship, and we will see why the so-called conflict thesis – the idea that science and Christianity are essentially opposed to each other – is mistaken. Here's how we will go about it:

Jews praying at the Western Wall, also called Wailing Wall, in the old city of Jerusalem

Dome of the Rock, Jerusalem

Mosaic of Jesus Christ, Hagia Sophia, Istanbul

Summary of this book

In Chapter 1, we look at the wonders and the benefits of science. We describe the universe as science understands it: from the vastness of time and space to the microscopic level, and then down to the unimaginably small world of quantum physics. We will also think about what we mean by 'science' and look at some of its history. We will see that, if science in the West was born in ancient Greece, it was nurtured in a Christian cradle before it gained its independence. And we will see that some people began to think that science should rebel and reject its Christian ancestry. Along the way in Chapter 1, you will be introduced to some high-flying scientists who are also unashamedly committed to the Christian faith.

Despite its power to reveal answers to many of our questions about the natural world, science has its limits. In Chapter 2, we will put science in a wider context, considering what sorts of questions lie outside the boundaries of science. We will see that some of the most important questions humans can ask, such as moral and existential questions, cannot be answered by science. We will also touch on the question of proof and certainty.

Chapter 3 introduces some light philosophy as we look at 'scientism', which is the ideology that arises if science is thought to be the only source of knowledge. This chapter goes a little deeper into the philosophy of science (see Article #11, "What is the philosophy of science?" on page 45), which concerns itself with how we gain scientific knowledge and what the limits of that knowledge are. We've kept more technical language like 'scientism' or 'methodological naturalism' for this chapter.

In Chapter 4, we will tackle 'the conflict thesis' head on. We will consider some of its history, and we will see why there is no insurmountable conflict between science and Christian belief. The short answer to how faith and science can get along is that they concern themselves with different sorts of questions and, for the most part, their truth claims do not conflict; rather, they complement one another.

The Conclusion wraps up the book and makes some suggestions about resources for thinking more about the important issues raised here.

Interspersed throughout this book you will also find various articles. They introduce you to a number of issues relevant to the book and they feature various scientists who are or were convinced Christian believers. These articles can be read separately, but they are also referred to in the main text.

Aristotle

Questions for discussion

Do you think there is a conflict between science and Christian belief?

Why might people say there is a conflict? Where does the conflict lie?

Could the idea of conflict be used as an excuse so as not to consider religious claims seriously?

"The heavens are telling the glory of God", says the psalmist in Psalm 19. Does pondering the wonders of the physical universe lead you to a sense of the presence or existence of God?

In his letter to the Romans, Chapter 1 verse 20, the Apostle Paul says that "ever since the creation of the world his [God's] eternal power and divine nature, invisible though they are, have been understood and seen through the things he has made". This seems to imply that the natural world should draw people to know something of God. What do you think Paul meant?

How would you explain to someone why some types of writing (genres) need to be read in a different way to other types of writing? (For example, a maths textbook, a novel and one of the Gospels.)

The Bible is not science: St Augustine's warning about confusing science and faith

Almost 1600 years ago, St Augustine of Hippo, the famous north-African bishop and theologian, warned Christians about the danger of confusing science and biblical truth. Science as we know it today did not exist in Augustine's time, but his warning is still relevant. He was concerned that Christians might look foolish if they misread the Bible by thinking that it could be used to make 'scientific' statements. In a commentary on the book of Genesis, he strongly warned Christians not to misuse the Bible by making supposedly biblical statements contradicting the knowledge of what we would call 'science' today. Here is what he said:

> Usually, even a non-Christian knows something about the earth, the heavens, and the other elements of this world, about the motion and orbit of the stars and even their size and relative positions, about the predictable eclipses of the sun and moon, the cycles of the years and the seasons, about the kinds of animals, shrubs, stones, and so forth, and this knowledge he holds to as being certain from reason and experience.
>
> Now, it is a disgraceful and dangerous thing for a non-Christian to hear a Christian, presumably giving the meaning of Holy Scripture, talking nonsense on these topics; and we should take all means to prevent such an embarrassing situation, in which people show up vast ignorance in a Christian and laugh it to scorn. The shame is not so much that an ignorant individual is derided, but that people outside the household of the faith think our sacred writers held such opinions, and, to the great loss of those for whose salvation we toil, the writers of our Scripture are criticized and rejected as unlearned men.

> **"Usually, even a non-Christian knows something about the earth, the heavens, and the other elements of this world."**

> If they find a Christian mistaken in a field which they themselves know well and hear him maintaining his foolish opinions about our books, how are they going to believe those books in matters concerning the resurrection of the dead, the hope of eternal life, and the kingdom of heaven, when they think their pages are full of falsehoods on facts which they themselves have learnt from experience and the light of reason?
>
> Reckless and incompetent expounders of Holy Scripture bring untold trouble and sorrow on their wiser brethren when they are caught in one of their mischievous false opinions and are taken to task by those who are not bound by the authority of our sacred books. For then, to defend their utterly foolish and obviously untrue statements, they will try to call upon Holy Scripture for proof and even recite from memory many passages which they think support their position, although "they understand neither what they say nor the things about which they make assertion".[1]

St Augustine of Hippo

Watch your genre! The importance of understanding text type

The warning of St Augustine in Article #1 highlights the importance of understanding what sort of book the Bible is. Whenever we read anything, we have an expectation of what sort of text it is. A manual for fixing a car should help us get back on the road, but it does not claim to solve existential questions. A play such as Shakespeare's *Macbeth* is about human relations and power and intrigue, but it is not meant to be a true telling of historical occurrences.

So, too, there is a fundamental difference between the Bible and a scientific textbook that tells us about the makeup and mechanisms of the natural world. As St Augustine noted so long ago, the Bible deals with "matters concerning the resurrection of the dead, the hope of eternal life, and the kingdom of heaven" (and much more), but it is not the purpose of the Bible to comment on the scientific aspects of "the motion and orbit of the stars … eclipses … the seasons … animals, shrubs, stones, and so forth".

The technical word for different types of literature is 'genre',

> **There is a fundamental difference between the Bible and a scientific textbook.**

which roughly translates as 'text type'. The Book of Genesis in the Bible and Darwin's *On the Origin of Species*, for example, are different genres. A vital question when reading anything at all is to ask what its text type or genre is. If you forget that Harry Potter is fantasy and not science, then you may find yourself in deep trouble as you try and make

Under the Milky Way, Great Ocean Road, Victoria

> **One last thing about the Bible that is important to remember is that it is made up of numerous individual books of many different genres written over many centuries.**

magic potions in the laboratory or leap off tall buildings on a single broomstick.

Some Christians take a literalistic approach to the Bible, perhaps thinking that "if the Bible says it I believe it". The problem is that this approach does not recognise the essential role of interpretation in all reading. As well, the literalistic view is not consistent. Many literalists will say that Genesis Chapter 1 speaks of six 'days' of creation, so God created the universe in something like 144 hours. However, the same literalist is unlikely to read Jesus' words "I am the true vine" (John 15:1) and take those words literally. Once again, the lesson is clear: understanding any text requires the interpretation and understanding of its genre.

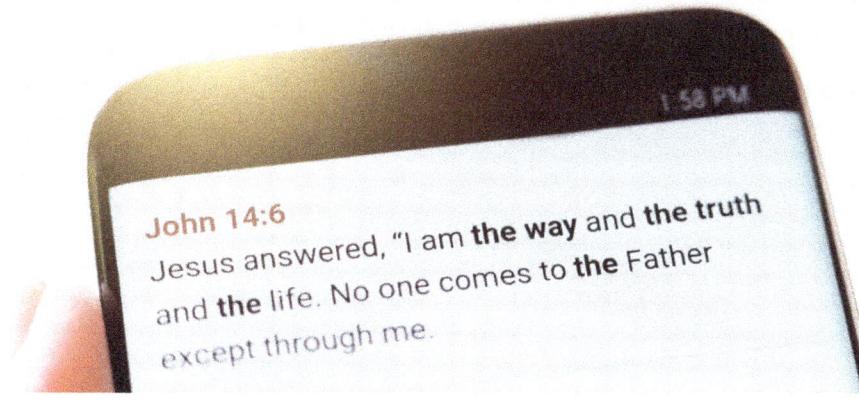

One last thing about the Bible that is important to remember is that it is made up of numerous individual books of many different genres written over many centuries. For example, the gospels are relatively straightforward accounts of the life of Jesus, but the book of Revelation (the last book in the Bible) is a highly symbolic and figurative depiction of the battle between good and evil and how God is ultimately in control of history. In the Bible, we find poetry and psalms and wisdom literature, for example, and all of these different text types need to be interpreted differently according to their genre. As well, while much is known about how the latter part of the Bible (the New Testament) came into existence, less is known about how the Old Testament came to be compiled from various documents and oral histories.

Charles Darwin's book, *On the Origins of Species by Means of Natural Selection*

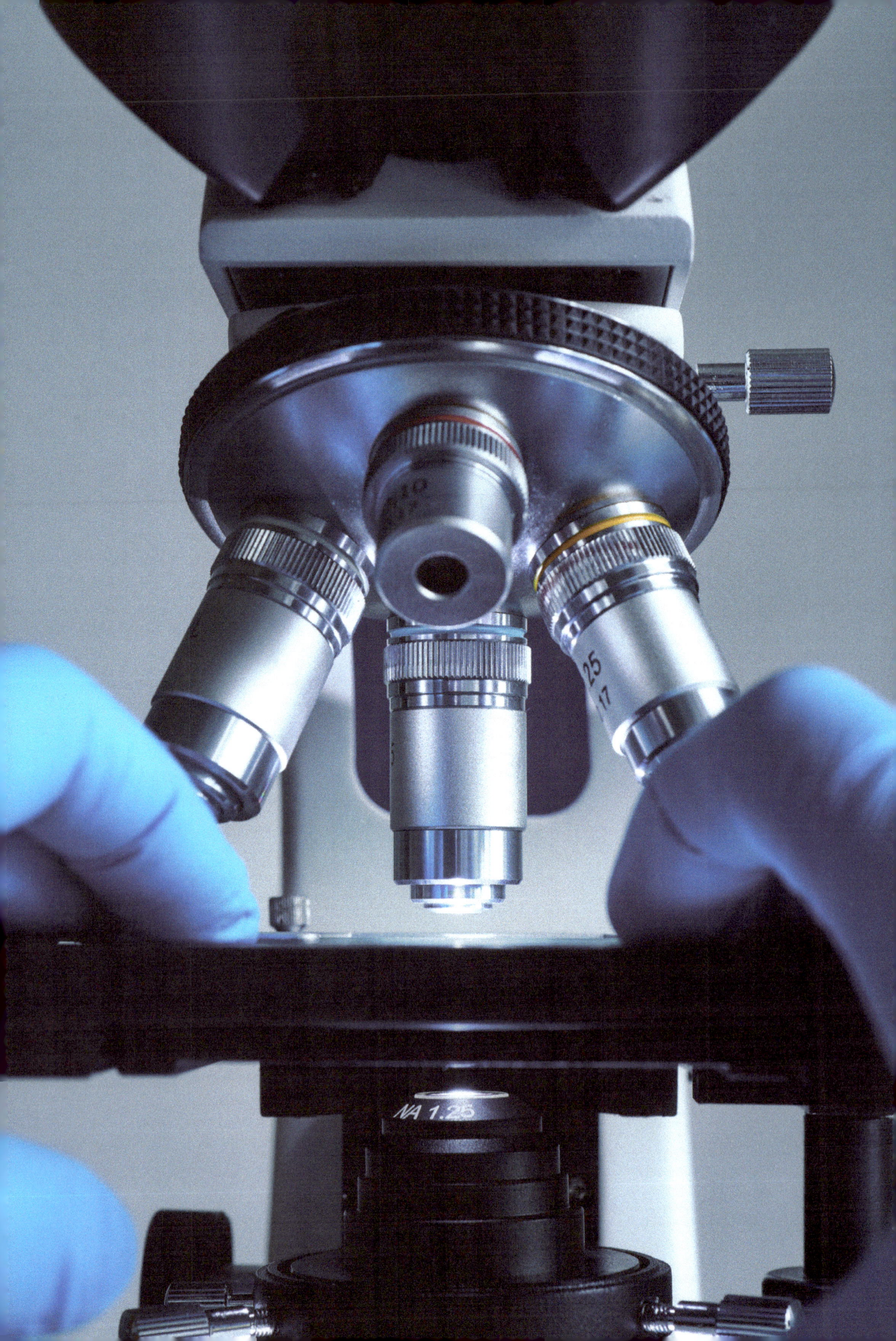

CHAPTER ONE

The wonders and benefits of science

For thousands of years, human beings have tried to make sense of their world in a way that animals probably don't. We ask questions about the world – about how things work, about causes and effects, about meanings and purposes. We look for explanations of the things that we see through telescopes and microscopes; we ask questions about good and evil; we ask questions about whether there is life after death and whether there is a purpose to human existence. And we also look for ways to make our world a better place by solving some of the practical difficulties of life on planet Earth.

So, this book follows in the tradition of strongly affirming the findings and the value of science. Science, with its possibilities for good, is one way of using our God-given abilities; it is a gift of God to humanity.

Looking at the Milky Way

Scientists search for answers to some of the questions on the previous page, while philosophers and theologians wrestle with other questions about meaning and purpose. In simple terms, the questions that science concerns itself with relate to the physical universe or the 'natural' world. And, as scientific knowledge continues to grow at an ever-increasing rate, it uncovers an awe-inspiring universe, ranging from the very, very small, subatomic level to the enormously vast cosmos of billions of stars and galaxies.

In this chapter, we will first look at some of the wonders and benefits of what we know as science today. Then we will consider some of the history of science before looking a little more at what we mean when we refer to science. Later, in Chapter 2, we will consider the limits of science, asking what sort of questions lie outside the realm of scientific knowledge.

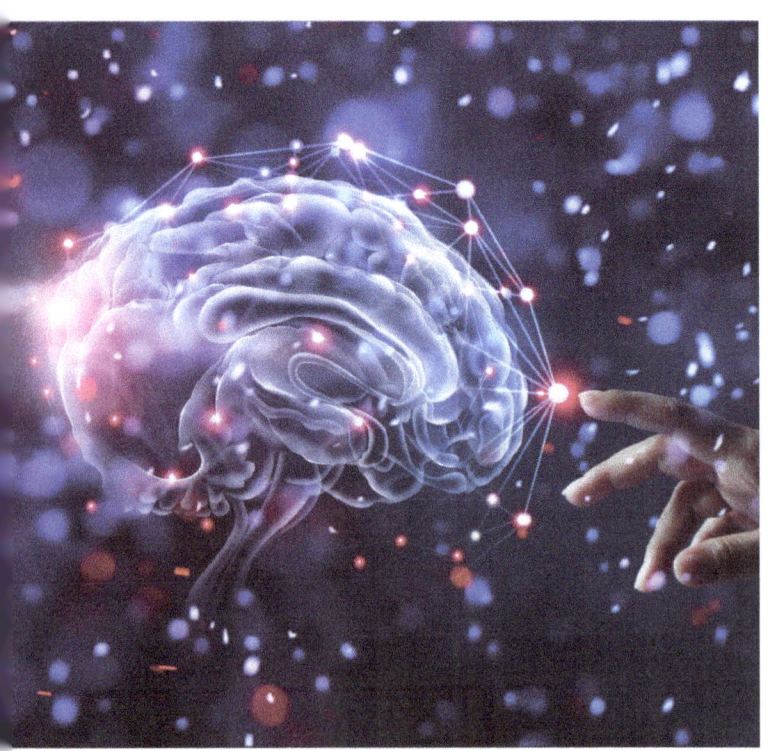

The surprises and benefits of science

In the introduction to this book, we heard the psalmist proclaiming the wonders of God seen in the creation. For many people, the decision to follow scientific careers is motivated by curiosity and wonder about the beauty, complexity and order in the natural world. For others, the power of science and technology to benefit humanity and the planet is a driving force. And, for many Christians, science is simply one area where they can fulfil what Jesus identified as the greatest of the commandments, to "love the Lord your God … with all your mind" (Matt 22:37).

Here is a taste of some of the extraordinary findings of science:

→ Did you know that light from the most distant parts of the universe has taken over 13 billion years (that's 13 thousand million years) to reach Earth? When we look into space, with or without a telescope, we are travelling back in time, seeing things as they were millions and thousands of millions of years ago. (See Article #6, "The Hubble Space Telescope: Looking back in time" on page 29 and Article #10, "Dr Jennifer Wiseman: Astrophysicist and public speaker" on page 30.)

→ Did you know that if you could unravel all the genetic code (DNA) in just one of your cells, it would be over two metres long? And if you add the DNA in all of a person's body end to end it would stretch from Earth to the Sun and back about 70 times. (See Article #8, "DNA: The instructions for life" on page 31 and Article #15 about Dr Francis Collins on page 58).

Dinosaur fossil, Australia

- Did you know that if the age of the Earth (about four billion years) was compressed into one year, the dinosaurs would have died out on about Christmas day, Jesus would have been born about 15 seconds before midnight on the last day of the year, and people alive today would live all their lives in the last second of that year? To put that another way, if the age of the Earth was the length of an Olympic marathon, all your life would fit in the last millimetre of that 42-kilometre distance. And the 2000 years since the birth of Jesus would be just 15 millimetres of the marathon. (See Article #4, "Monsignor Georges Lemaître: Priest and father of the Big Bang" on page 20.)

- Did you know that even though we know the 'law' of gravity so well that we can land people on the moon, we still don't know why things fall down? As Albert Einstein would have said, we don't know why matter distorts space-time. (See Article #4, "Johannes Kepler: Thinking God's Thoughts After Him" on page 20.)

- Did you know that there are more stars in the universe than grains of sand on all the beaches and deserts of our planet?

- Did you know that there are more atoms in the human body than stars in the universe? And did you know that the atoms that make up your body were actually formed in stars millions of years ago? We are made up of stardust!

- Did you know that millions of your body's cells, mostly blood cells, die every second? Fortunately, millions of new cells form every second too.

- Did you know that two particles billions of kilometres apart can be 'entangled' so that changing the properties of one will change the properties of the other instantaneously? (See Article #17 about Sir John Polkinghorne on page 60, and Article #3, "The weird, weird world of quantum physics" on page 18.)

Radio telescopes observe the Milky Way

- Did you know that the most complex thing in the universe, the soft, squelchy human brain, has about 100 billion neurons (cells), each one connected to hundreds or thousands of others? In the first three months of pregnancy, 250,000 neurons are formed every minute. In an adult, almost a litre of blood passes through the brain's 160,000 km of blood vessels every minute; without blood flow unconsciousness occurs within seconds and the brain will die within minutes.

What an extraordinary universe science reveals!

Not only does science reveal an amazing world and satisfy our longing for knowledge; it also brings great benefits when this knowledge is turned to practical applications.

- Without science we would not have modern medicine and drugs to combat disease and pain. (See Article #4 about Professor Graeme Clark and the bionic ear on page 19.)

- Without science we would not have modern agriculture, enabling us to produce nutritious and abundant food. (See Article #9 about Professor Phil Batterham and crop protection on page 9.)

- Without science we would not have the ability to speak to people further away from us than the length of a football field.

Lakebed damage from drought

- Without science we would not know about the dangers of climate change or how to mitigate it or adapt to it. (See Article #10, "Climate science and Christian belief" on page 33.)
- Without science we would not be able to travel further or faster than a horse could take us.

The history of science

Defining 'science'

Have you noticed that the phrase "science as we know it today" has appeared a number of times in this book? Yes, there is a problem with using the word 'science' because what we think of as science today (which is also hard to define exactly) is certainly not an age-old idea. Both the word 'science' and the practice of what we call science have changed dramatically through history. It's important to be aware of this danger of importing current ideas into our thinking about historical people and practices. Although we sometimes talk of science and scientists of 500 or 1000 or more years ago, those people were not doing what we think of as science now.

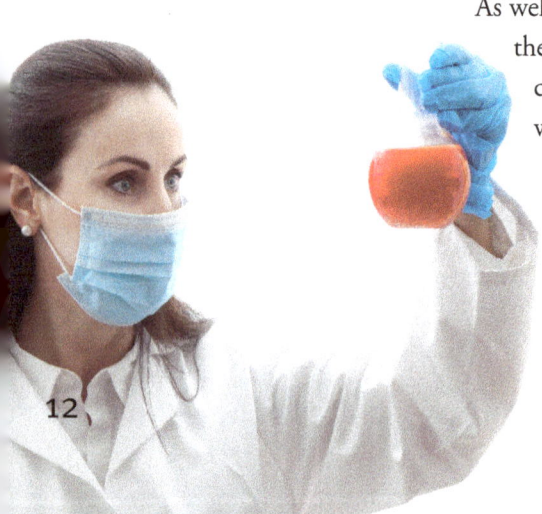

As well as being aware of the newness of our concept of science, we should also be aware that rightly speaking, we should talk of 'sciences' not 'science'. Although this book uses the singular term 'science', in fact there are numerous sciences, each one with its own slightly different methods. For example, the methods of physics, which are highly mathematical, are not the same as the methods of evolutionary biology. Nevertheless, most sciences are based on a process of asking questions, making observations and measurements, proposing hypotheses and attempting to verify or falsify them.

The history of science as we know it in the West today is a long, gradual story with its origins in Greek thinking from about 2500 years ago. It was then that people first tried to explain the world around them in terms of natural causes and reasons rather than in unscrutinised myths and stories. Rather than seeing the cosmos (Greek for 'world' or 'universe') as chaotic and unpredictable, Greek philosophers began to think of it as behaving consistently. If this were the case then, they thought, it should be possible to use human reasoning to understand the universe. This was the beginning of what was called 'natural philosophy', the forerunner of science. However, it took many centuries before the so-called scientific revolution occurred and the emergence of the rigorous pursuit we call science today. And it was only about 150 years ago that William Whewell (a Cambridge academic and Anglican minister) invented the word 'scientist' to describe those who did the work of science.

Rev. Dr William Whewell

> **One way of understanding the amazing power and flourishing of science is to think of it as a rope made up of a number of strands.**

The roots of the word 'science' lie in various languages and point to the idea of knowledge. However, science as it is understood today usually refers to an organised examination of the natural world, based on observation and, where possible, on experiment. The *Oxford English Dictionary* puts it like this: "Science is the intellectual and practical activity encompassing the systematic study of the structure and behaviour of the physical and natural world through observation and experiment." Science results in generalisations helping us understand the workings of nature, and it aims to make predictions that can be tested.

According to Dr Francis Collins (see Article #15, "Dr Francis Collins: Medical researcher and genetic pioneer" on page 58):

> Science is progressive and self-correcting: no significantly erroneous conclusions or false hypotheses can be sustained for long, as newer observations will ultimately knock down incorrect constructs. But over a long period of time, a consistent set of observations sometimes emerges that leads to a new framework of understanding. That framework is then given a much more substantive description, and is called a 'theory' – the theory of gravitation, the theory of relativity, or the germ theory, for instance.[2]

One way of understanding the amazing power and flourishing of science is to think of it as a rope made up of a number of strands. Science gains its effectiveness from the interweaving of those strands that took centuries to eventually come together a few hundred years ago in what we call 'the scientific revolution'.

One strand of science is natural history. The naturalist or natural historian (for example, Charles Darwin, the 'father' of the theory of evolution), is adept at rigorous observation, description and cataloguing of the natural world. The Greek philosopher Aristotle, in the fourth century BC, was the first to study living things in a systematic way, and he could lay claim to be the first natural

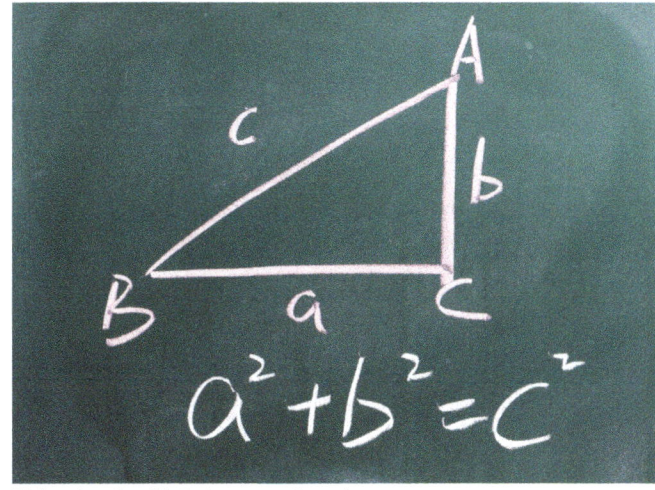

Right triangle with Pythagorean formula

historian. Although we don't have any 'field diaries' that note his on-the-ground observations, we do have his written-up biological notes, perhaps akin to a textbook today. Another very early naturalist (just one of her many gifts) was the Benedictine abbess St Hildegard of Bingen, who lived in the twelfth century and is considered the founder of natural history in Germany (see Article #4, "St Hildegard of Bingen" on page 19).

Another of the strands of science is mathematics. Mathematics was used around 5000 years ago by the Babylonians and Egyptians for astronomy, as well as for construction and financial calculations. Later, around 2500 years ago, Greek mathematicians such as Pythagoras, Euclid and Archimedes formalised mathematics as a field of study. However, it was in the time of the scientific revolution that mathematics was consistently applied to the physics of the day (then called natural philosophy). In 1623, the famous astronomer and Christian Galileo Galilei (see Article #14, "Clearing the air about the Galileo affair" on page 56) said that the universe was written in the language of mathematics and that it could not be understood without maths:

> The universe … cannot be understood unless one first learns to comprehend the language and read the letters in which it is composed. It is written in the language of mathematics, and its characters are triangles, circles, and other geometric figures without which it is humanly impossible to understand a single word of it; without these, one wanders about in a dark labyrinth.[3]

Another strand of science is experiment. Science is constantly testing its theories, and the way to test many scientific theories (but by no means all) is to devise an experimental test. It works like this: if you have a theory – for example, about the laws of gravity and how fast a falling object accelerates – then the theory can be tested by working out a way of taking lots of measurements and seeing how they compare to what your theory tells you should happen. If the measurements line up with the theory, then they are said to confirm (but not prove) the theory. If the measurements do not line up with the theory, perhaps the theory is wrong, or perhaps there was an error in making the measurements. Later in this chapter, we will see why a theory can never be finally proven beyond any doubt, even if all the measurements confirm the theory.

Another less obvious but essential strand of the scientific enterprise is a set of beliefs or assumptions about the world and about human interaction in the world. In a sense, these assumptions or presuppositions are not so much part of the practice of science as they are the foundations on which science is built. For example, scientists believe that human beings are capable of grasping the nature of the world around us, and they believe that we live in an ordered universe and not a chaotic one. We will return to look more at these foundational assumptions of science, but first let's consider a little more history.

The two books of God and an ancient warning

One common and long-used way of describing the harmony between science and Christian belief is to refer to the 'two books of God': the book of God's works (the creation or the natural world) and the book of God's word (the Bible). Sometimes these 'books' are referred to as the book of Scripture and the book of nature. It is interesting that perhaps the most famous book associated with science–faith conflict, which is Charles Darwin's *On the Origin of Species*, affirmed the role of the Creator and promoted the two-books tradition (see Article #5, "Charles Darwin affirms God's two books" on page 21).

This image of God's two books affirms that both are worthy of study in order to arrive at truth. However, as we saw in the introduction, St Augustine warned, some 1600 years ago, that Christians should be wary about overstepping the mark by using the Bible to assert scientific truths.

Another common phrase used by Christians when thinking about the breadth of human knowledge, following St Augustine and St Thomas Aquinas, is that "all truth is God's truth". This implies that Christians do not need to fear an honest search for truth. If "all truth is God's truth" then wherever truth is revealed it is the truth of God's world that is being uncovered (see Article #4, "Johannes Kepler: Thinking God's thoughts after him" on page 20). This understanding, along with Jesus' call, mentioned above, to "love the Lord your God … with all your mind" (Matt 22:37) has led many committed Christians into intellectual and scientific vocations in the pursuit of truth. For those who dedicate themselves to science, that truth is about the natural world, which is also God's creation. For these people, there is no fundamental conflict between science and Christian belief. Indeed, each supports the other.

It is only in the last two or three centuries that there has been the suggestion that an insurmountable conflict exists between Christianity and the pursuit of science. In recent decades this idea of a fundamental science–faith conflict has been promoted by the so-called New Atheists, the best known of whom is perhaps Professor Richard Dawkins. The New Atheists tend to promote the idea that science is the only and fundamental source of truth. We will return to this view, called 'scientism', in Chapter 3, and then in Chapter 4 we will examine the conflict thesis more closely and show why a harmony between science and Christianity is possible.

The presuppositions of science

The life and breath of science lies in its rigorous approach to uncovering the truth of the natural world. However, to understand science clearly, it is important to note that science is founded on certain working assumptions, which it does not question. This recognition that science doesn't start from a blank slate – that science must assume some things to even get off the ground – is captured by atheist philosopher Daniel Dennett, who warns against the danger of a naive attitude to science, which fails to see its philosophical foundations: "There is no such thing as philosophy-free science; there is only science whose philosophical baggage is taken on board without examination."[4]

One way of thinking about these underpinning assumptions of science is that they are like tools of the trade that we use to produce results. The carpenter uses a hammer without questioning it in order to drive a nail; the focus is on the nail while the hammer is taken for granted. So, too, science takes for granted its foundational assumptions. However, that doesn't mean that they are proven. Science cannot justify its assumptions scientifically; they must come first before science even begins its work. If we don't accept these assumptions, then we would never be confident about the claims of science.

So what are some of the presuppositions of science?

Scientists assume that we live in an ordered world

In order to conduct scientific investigations, the scientist needs to work on the assumption that the world is an ordered, regular one and not a chaotic one. For example, we assume that gravity follows the same 'laws' every day and everywhere; gravity doesn't turn on and off randomly. If the world were chaotic, then there would be no consistent reasons for things to happen as they do. However, if the world is ordered, we can search for that order, uncovering what we usually refer to as the 'laws of nature', which describe the order that we encounter. And, assuming there are laws governing the natural world, we can design experiments that can be repeated, expecting to get the same results each time. As a result, in the laboratory, the scientist assumes that the results of an experiment are due to the laws of nature and not to either random or miraculous causes. This assumption about the ordered nature of the universe governs the scientist's methods of going about science, and it is an assumption that cannot be ultimately proven.

Scientists assume that the world could be different

Imagine that there is only one possible world in which everything in the world has to be exactly the way it is.

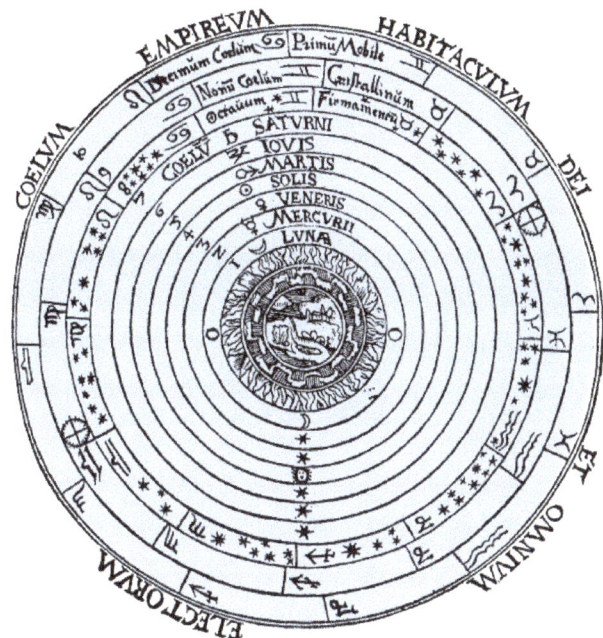

The Ptolemaic system with Earth at the centre of the universe, Peter Apian, *Cosmographia*, Antwerp 1524

Imagine that the universe cannot possibly have one more or one less star in it. Imagine that the universe could not possibly be ordered in a different way to the way we find it. Such a universe would be a 'necessary' universe; it could not be different to the way it is. Just as a triangle must have only three sides, so, according to this thinking, the universe must be the way it is.

For many centuries it was assumed that the world had to be the way it is because of some sort of necessary logical truths (like a triangle must have three sides). Aristotle thought the circle and sphere were perfect shapes so therefore the universe itself must be a series of concentric spheres with all the stars and planets moving in perfect circles. This 'Aristotelean cosmos' was the accepted view for almost 2000 years.

Science, on the other hand, is based on the assumption that the world does not necessarily have to be the way it is. For example, gravity could be a stronger or a weaker force than it is. And, if we believe that things might have been ordered differently to the way they are – according to different laws perhaps – then the only way to find out how they are actually ordered is to investigate the world.

The technical way of expressing this 'possibility of being different' is to say that the natural world is 'contingent'. Contingency ('things could be different to how they are') is the opposite of necessity ('things must be the way they are'), and scientific investigation into how the world is assumes the world could be different to the way it is.

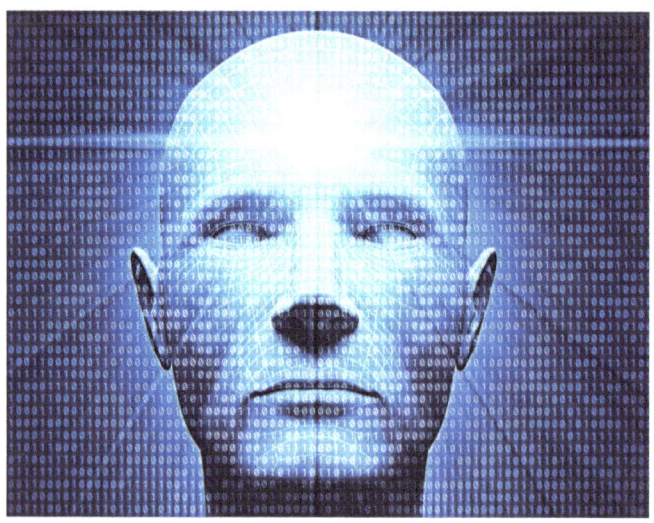

A 3D head in binary code

Scientists assume that we are able to understand the world

Have you seen *The Matrix*? In this movie, people who think they experience and understand the real world are actually in the 'matrix', plugged into computers that make them think they are living in the 'real' world. How would we know if we were in the matrix? How do we know that humans have the ability to make contact with the real world and understand it? It's not surprising if these seem like odd questions because we think it is obvious that we are in touch with reality. However, the experience of a vivid dream or hallucination might make us realise that the question is not so easy to answer.

Science also assumes that there is a world 'out there', which is independent of whatever human beings might think or say about it. It is notoriously difficult to rigorously prove the existence of what philosophers call the external world – the world outside our minds. It is simply something we accept as true without question, and it seems absurd to demand proofs for what we take to be so obviously true. Nevertheless, scientists assume that we are able to make meaningful contact with that reality, which ranges from subatomic particles to galaxies billions of light years away, and from the very tangible (like your hand in front of your face) to the very intangible (like dark matter and energy, or magnetic fields).

As well as taking for granted the existence of the external world, science must also assume that human reasoning and memory and sensory functions are reliably sound and lead to true understandings of that natural world. Again, we cannot prove these assumptions because they are assumptions we must make in order to even think about any sort of proofs or argument.

In addition to assuming that there is a world outside our minds and that we can grasp it, there are other assumptions that ground scientific discussion. Human reasoning itself depends on the concept of truth and the validity of basic rules of logic, which are also assumptions we must make before we can begin a rational conversation. You can't argue for the reliability of logic without using logic. So, for example, we must take for granted that you cannot assert one thing and its opposite without falling into incoherence. Either all swans are white or they are not, but you can't have it both ways, and if you think you can then you leave yourself out of rational conversation.

Scientists assume that inductive reasoning is effective

One type of reasoning that science depends on is called induction. Observations about the world are the basis of most scientific theories. Much of science involves drawing conclusions or framing hypotheses based on the regularities of the data that we find in the world around us. In a very general sense, scientific thinking moves from particular observations to general conclusions. This process of reasoning from particular cases to general conclusions is an example of induction, or an inductive argument. Inductive argument is the process of observing patterns or repeated events or experience or experimental results, and drawing general conclusions on the assumption that future or unobservable events and instances will follow the same pattern.

More generally, induction involves drawing a conclusion that seems probable, based on the evidence, but it is not certain. For example, if we observe that many swans are white, we might wonder if all swans are white. After observing many swans, if we never see a black swan, then we might hypothesise or draw the conclusion that "All swans are white". In this case, we have made an 'inductive inference' from *some* cases to *all* cases. In this example of the inductive process, we work from particular examples (*this* swan, *that* swan,

Black swan

One type of reasoning that science depends on is called induction. Observations about the world are the basis of most scientific theories.

that other swan …) to a general rule (*all* swans). In fact, Europeans believed that all swans were white until, in 1697, the Dutch explorer Willem Hesselsz de Vlamingh found black swans in what is now Perth, Western Australia.

The reason that science needs to use induction is that the only concrete information that we have about the universe is the data that we observe piece by piece. As we amass information, we use induction to form general conclusions (usually called hypotheses and then, as they become confirmed, theories) that eventually come to be called scientific laws. For example, Isaac Newton formulated the universal law of gravitation[5] based on observations of the behaviour of numerous objects such as the moon and infamously, but perhaps only anecdotally, a falling apple.

Blue mandala

Conclusion

In this chapter we have seen that science is awe-inspiring and that science is a constant part of human flourishing: it offers great benefits to humanity and life on this planet. We have also seen that science does not work in a philosophical vacuum, so to speak. That is, we have seen that science rests on certain presuppositions or assumptions that need to be made in order to get science off the ground.

The implications of this are that, while science is a rigorous pursuit of the truth about the natural world, it is also subject to certain assumptions (as well as human and social factors), which means that scientific claims can be described in the same way as legal claims in terms of degrees of confidence and being 'beyond reasonable doubt'.

We have also seen hints, which will be followed up in the rest of this book, that science and Christianity have traditionally been friends.

In the next chapter, we will examine the boundaries of the scientific enterprise, asking what the limits of science might be. We will see that, while science is enormously fruitful in answering certain types of questions, there are other sorts of questions that science just cannot deal with.

Questions for discussion

What do you find interesting or amazing about science and scientific discoveries?

What questions are raised in your mind when you hear extraordinary scientific facts?

What do you think are the greatest benefits of science?

What do you think are the biggest problems raised by science?

What sorts of questions do you think lie outside the boundaries of science? In other words, what questions can't science answer?

Why might a religious person dedicate their life to a scientific career?

The weird, weird world of quantum physics

Quantum physics is the science of the very, very small, and the quantum world is very, very weird.

One odd characteristic of the quantum world is that it is difficult to describe in normal language what quantum things are like. Often we refer to them as particles – imagine a ping pong ball or a dust speck – but they can also be described as 'waves'. Imagine waves in the ocean where the wave moves across the water, but the individual water droplets simply move up and down. Subatomic matter sometimes behaves like a particle and sometimes like a wave. And although scientists can predict where these quantum bits might be, we can never know for certain.

Perhaps the most famous example of quantum weirdness is a thought experiment proposed by Erwin Schrödinger in 1935. Schrödinger was a physicist trying to explain the behaviour of photons, which are the subatomic 'packets' of light and radiation. Photons, like other bits of atomic stuff, can behave as waves or particles, depending on the experiment, but it is never possible to see them behaving as both. This means that whether a photon is a particle or a wave is only decided by the experiment. In other words, the observer 'forces' the photon to be a particle or a wave, but until that time, the photon is not one or the other. Or it is both! (Scientists call this 'complementarity'.)

Schrödinger described an experiment about a cat to highlight the perplexing problem of the quantum world being dependent on whether someone is watching or not. Imagine a box with a cat in it and a device that could release poison gas. Now imagine an automatic triggering device in the box that gives equal possibilities to whether the gas would be released or not (the same chance as flipping a coin). Therefore there is a 50 per cent chance the cat will die, but the only way to find out if the cat is dead or alive is to look inside the box.

Schrödinger said that if the device in the box behaved like photons, which can be a particle or a wave, then we would have to think of the cat as both dead and alive until the moment of opening the box. Then, the cat's fate is determined by the observer who performs the experiment.

Imagine a box with a cat in it and a device that could release poison gas.

Another counter-intuitive aspect of quantum physics is called 'entanglement', or what the famous physicist Albert Einstein called "spooky action at a distance". Quantum entanglement refers to the fact that two subatomic particles can be linked together across great distances so that if you change one particle the other one will instantly change too. Experiments have so far demonstrated this sort of entanglement working at over 1000 kilometres.

Diagram of Schrödinger's cat thought experiment

Visionary figures in science

Graeme Clark: Bionic ear pioneer and a man of prayer

Professor Graeme Clark is one of Australia's most awarded scientists, receiving numerous international science prizes for achieving his dream of allowing deaf children to hear. When he started his work decades ago, it was said that whoever succeeded in the task would surely win a Nobel Prize. Although it might still be a possibility, Professor Clark says that doing work to win prizes is not what he set out to do. "My aim was to help people and do God's will," he says. "But thanks to God to have given me some talents … the prize for me is seeing the children, who would not have been able to speak, communicating with their parents and friends. It brings tears to my eyes."[6]

In his life as Australia's bionic ear pioneer, Professor Clark is an example of the definition of faith found in Hebrews 11:1: "Now faith is the assurance of things hoped for, the conviction of things not seen." For Professor Clark, science and Christianity share this in common: they are both exercises in faith because they both involve a resolute "conviction of things not yet seen", as well as a confidence of arriving at the goal. It might be the Christian conviction that one day God will "make all things new", or it might be the strong belief that one day children born deaf will be able to hear. Either way, it is a case of perseverance: "Everyone said it wouldn't work and that I was foolish," Professor Clark says. "It was only some years afterwards when it was finally successful that I became accepted as a scientist."

Professor Clark describes this decades-long perseverance in both his science and his faith as a journey and a calling: "It has been a deep calling. I have learned to be faithful in prayer and in my scientific work, and I've experienced in my journey that the two are interrelated. The spiritual side for me has been fundamental."

Professor Clark speaks of prayer and the presence of God as being part of his life's journey:

> I could not rely on myself alone when taking on this journey developing a bionic ear. It was so demanding, so traumatic, and most people said it wouldn't work; people said I was a clown. My colleagues sent word to the vice-chancellor that I should be relieved of my job … these pressures while trying to do something that was breaking new ground – that's why it was a matter of really constant prayer.

St Hildegard of Bingen: Abbess and founder of German natural history

St Hildegard of Bingen (1098–1179) was an abbess who lived almost 1000 years ago in what is now Germany. She was a true polymath and, if we were forced to use modern terms, we might describe her as a scientist, a theologian, a composer, a writer and a preacher. Apart from her theological and mystical works, she also wrote about medical diseases and cures, and she catalogued observations of plants and animals. Consequently, she is considered to be the founder of scientific natural history in Germany.

St Hildegard of Bingen

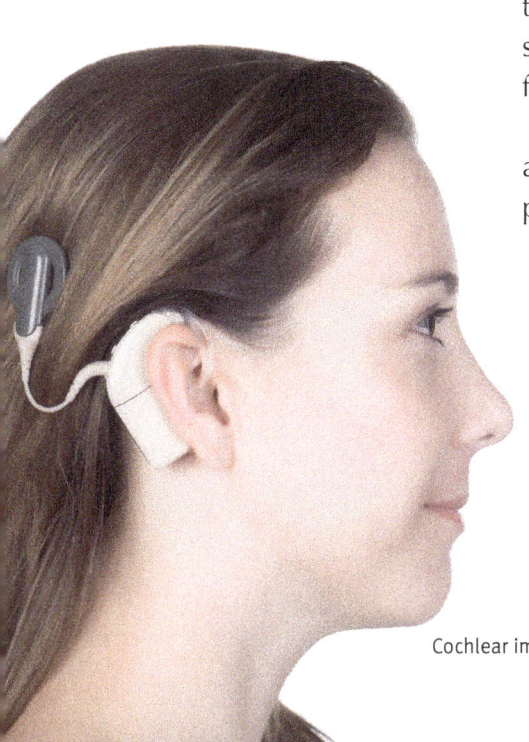

Cochlear implant

The Big Bang beginnings

Johannes Kepler: Thinking God's thoughts after him

Johannes Kepler (1571–1630) – astronomer, mathematician and theologian – was a contemporary of Galileo, and one of the giants of the scientific revolution. He is best known for his three laws of planetary motion which overturned the idea that the planets must travel in circular orbits. He described mathematically how the planets could be moving around the sun in an elliptical fashion. Kepler was set to become a Lutheran priest in his native Germany, but he turned to science for deeply theological reasons. Later, he is reputed to have described his scientific investigations as revealing the glory of the Creator in this way:

> I was merely thinking God's thoughts after him. Since we astronomers are priests of the highest God in regard to the book of nature, it benefits us to be thoughtful, not of the glory of our minds, but rather, above all else, of the glory of God.[7]

Monsignor Georges Lemaître: Priest and father of the Big Bang

Georges Lemaître (1894–1966) was a Belgian priest and scientist, and he was a forerunner in describing the nature of the universe as we now understand it. In 1927, after observations and theoretical calculations showing that the universe was expanding, Lemaître proposed that this expansion could be extrapolated backwards in time to an initial point. He called it the hypothesis of the primeval atom, which later became known as the Big Bang theory after astronomer Fred Hoyle used the term to describe Lemaître's competing view to his own preferred one, which was the steady state 'eternal' view of the cosmos.

While Lemaître was in no doubt about the harmony between science and his Christian belief, he did not believe that his scientific theory should be used for religious purposes. "As far as I can see, such a theory remains entirely outside any metaphysical or religious question," he said.[8]

Although Lemaître was the first to publish the research calculating the rate of expansion of the universe, he was relatively unknown at the time, and the law came to be known as Hubble's Law, named after Edwin Hubble who, two years after Lemaître, published his own calculations.

Johannes Kepler

Charles Darwin affirms God's two books

On the Origin of Species by Charles Darwin was published in 1859 and popularised the evolutionary theory of natural selection. Its publication prompted one of the most famous examples of a so-called conflict between science and faith. However, it's interesting that Darwin did not seem to think there was such a conflict and, in fact, he specifically referred to God at the start and end of *Origin*. Opposite the title page of the first editions of the book, Darwin placed the words of Francis Bacon affirming the necessity of studying both of God's books:

> Let no man out of a weak conceit of sobriety, or an ill-applied moderation, think or maintain, that a man can search too far or be too well studied in the book of God's word, or in the book of God's works; divinity or philosophy; but rather let men endeavour an endless progress or proficience in both.[9]

Its publication prompted one of the most famous examples of a so-called conflict between science and faith.

Meanwhile, on the last page of *On the Origin of Species*, Darwin wrote the following (although the words in brackets do not appear in the first edition of the work):

> There is grandeur in this view of life, with its several powers, having been originally breathed [by the Creator] into a few forms or into one; and that, whilst this planet has gone cycling on according to the fixed law of gravity, from so simple a beginning endless forms most beautiful and most wonderful have been, and are being evolved.

Charles Darwin

These quotations from Darwin's classic work make it clear that the author of *On the Origin of Species* was not averse to setting evolution in a theistic framework. Although Darwin was not a traditional Christian, and was probably an agnostic by the end of his life, it is simply not true, now or when Darwin popularised the theory, to suggest that evolution was or is necessarily atheistic.

The theory of human evolution

CHAPTER TWO
The limits of science

In the previous chapter, we looked at the wonders and benefits of science, and we also saw that science is built on a foundation of assumptions or presuppositions that, in themselves, are not scientific in the sense that they do not arise from science. Rather, they must be assumed in order to do science.

In this chapter, we will consider the bounds of science: What sorts of pursuits lie within the legitimate realm of science? What sort of questions and activities lie outside science? In other words, what questions might science answer one day, and what questions will science never be able to answer?

Let's start with an illustration …

Why is the water boiling? Meanings and mechanisms

Imagine this: you go into the chemistry laboratory and the teacher, Dr Smith, already has a Bunsen burner set up under a beaker of water which is boiling vigorously. "Today we have a test … Why is the water boiling?" asks Dr Smith.

Little Jim, the smartest student in the class, shoots up his hand. "The water is boiling because the stored chemical energy in the gas is converted to heat energy and transferred through the glass to the molecules of water in the beaker. The molecules get hotter and start to move more and more violently until eventually, at about 100 degrees Celsius, bubbles of water vapour start forming and coming out of the liquid water. That's why the water is boiling," says Little Jim.

"Mmm, good try," replies Dr Smith, "but actually, the water is boiling because I haven't had my morning cup of tea." And, with that, she pulls a teabag out of her pocket and dips it in the beaker of boiling water.

So, which is the correct answer? Dr Smith's or Little Jim's? It's fairly easy to see that in this case they are both right. The water is boiling because of the heat transfer from the burning gas, and it is boiling because Dr Smith wants her morning tea. There are two correct answers to the question!

Both answers are correct. However, they show that the question, "Why is the water boiling?", is ambiguous – it can be understood in more than one way. Or, perhaps it is two questions in one. It could be a question about *mechanics*: "What causes the water to boil?" Or, it could be a question about *meanings*: "What is the purpose of the water boiling?" It could be a question about *particles* or it could be a question about *purposes*.

In other words, a question might be a scientific question or it might be a non-scientific question. There are lots of kinds of questions that are not scientific, as we shall see below. And, as this example illustrates, it's important to understand what sort of a question we are trying to answer.

Let's ask another question: "Why are human beings here on Earth?" This is the sort of question that has both a theological and a scientific answer. It could be answered by telling the story of evolution that finishes with human beings at the top of the tree of life. It can also be answered by a Christian speaking of the creation by God of the universe and Earth, and of life on this planet. As we will see, just like the answer to "Why is the water boiling?", the two answers to "Why are we here?" do not have to compete with each other.

Science has its limits

We have now clarified two types of questions and answers, which, as shorthand, we are referring to as questions about meanings and questions about mechanisms. And if, as we have suggested, science is mostly about mechanisms and not meanings, then we can see that there are limits to science imposed by the sorts of questions it can legitimately ask and the sorts of answers it can expect.

It may be helpful to think of two distinct categories of difficulty for science, which highlight these limitations. The first category includes the sort of questions that science does not answer yet, but that we might imagine science answering one day. The second group of questions are those that science will never explain because they lie outside its domain. Let's turn first to some of the difficult questions in science that may one day be resolved. After that we will look at some questions that science by its nature will never be able to answer. Let's call the first category the 'practical limits' of science and the second category, the 'philosophical limits' of science.

Practical limits of science

While science has been enormously successful, there are some so-far-unanswered questions that make us realise that we are very far from knowing everything about the natural world. Think of challenges such as the following:

Gravity. We understand the *laws* of gravity very well: well enough to land people on the moon and to keep satellites very accurately in orbit. However, the search for an explanation of gravity is still unsuccessful. A popular history of science subject at a Melbourne university ends the course with memorable words along these lines: "After 2500 years of searching for the answer, natural philosophers and scientists still don't know why things fall down."

"Why are we here?" This is the sort of question that has both a theological and a scientific answer. It could be answered by telling the story of evolution that finishes with human beings at the top of the tree of life. But it can also be answered by a Christian speaking of the creation by God of the universe and Earth and life on this planet.

Dark matter. According to current theories, most of the mass of the universe (about 85%) seems to be made up of 'dark matter', which cannot be seen because it does not reflect or emit light. We have no idea what type of matter it is, but its existence is postulated in order to explain observations of the movements of the known stars and galaxies and the expansion of the universe.

Fine-tuning. The laws of the universe appear to be fine-tuned for the existence of life. There are various mathematical constants in the universe that, if they were slightly different, would make life impossible. Is it a coincidence that we live in such a universe? The main proposed explanation extends the bounds of believability. It's called the multiverse theory and postulates that myriad parallel universes exist, which cover all the possible values of the fundamental constants. Physicist Paul Davies said this about the multiverse theory:

> Invoking an infinity of unseen universes to explain the unusual features of the one we do see is just as ad hoc as invoking an unseen Creator. The multiverse theory may be dressed up in scientific language, but in essence it requires the same leap of faith.[10]

The beginnings of life. While evolutionary theory postulates that all life is descended from an original life form, we seem as far as ever from seriously answering how a reproducing life form could have come into existence from non-life. The most common view is that there was a primordial 'soup' of organic compounds that gave rise to life when lightning or radiation added energy to the mix.

Consciousness. The struggle to understand the subjective aspects of consciousness has recently bridged from philosophical discussions to incorporating the findings of neuroscience. But, while correlations have been revealed between brain states and the feelings and attitudes that we attribute to consciousness, we are still no closer to understanding how any particular brain state can be mapped onto "what it is like to be me". As one respected physicist says, "it is the only major question in the sciences that we don't even know how to ask".[11]

Free will. While we all act as if we have free will, a 'strictly scientific' view of human beings seems to lead to the conclusion that everything that we do, say or think is ultimately determined by strict causal laws or by random sub-atomic events. All normal life, including science itself, assumes and depends on human freedom of choice, yet explaining how a strictly biological and chemical view of human beings is compatible with that freedom is scientifically and philosophically out of our reach at present.

These are some of the difficult questions that science wrestles with. Now let's turn from the difficult to the impossible questions; those issues that science can't hope to ever explain because they lie outside of science altogether.

Philosophical limits of science

Not only are there practical limits to science, some of which seem very difficult to overcome, but there are also philosophical limits; these are limits imposed on science by the very nature of science itself. As a pursuit of knowledge about the natural world, science cannot delve into philosophical or moral or logical or religious questions. Science can't do so because such questions are not the subject matter of science.

Science can't answer moral questions; it can't tell us right from wrong

The advance of science and technology raises many moral questions about how science should be put to use. For example, science contributes to increasingly lethal means of warfare. However, while science and technology can make weapons of mass destruction or autonomous attack drones and robots, science cannot tell us whether such weapons should be used or not. Another question: Should we legalise euthanasia? That's a question that science cannot answer, even though medical

Prof. Stephen Hawking

Prof. Richard Dawkins

developments can make it more easy and less painful to kill a human being. Science and technology have also given us many means to prolong life and to change our physical and genetic makeup. When does therapeutic medicine become unethical enhancement? How far should we go in making ourselves 'better'? And who defines what 'better' means?

These questions about what we 'should' do are moral questions. Should torture or capital punishment or abortion be allowable in some circumstances? Should our country accept more refugees? Should gene editing of embryos be allowed, and under what circumstances? These are all moral questions that science cannot answer because they lie outside the bounds of legitimate science.

These questions highlight the 'is–ought gap' between empirical facts ('is' questions) and moral claims ('ought' questions). Philosophers are prone to say that you can't get an 'ought' from an 'is'. So, while science is excellent at telling us how the world 'is', it can't tell us how it 'ought' to be. (See Article #8 on page 31 which discusses some moral questions associated with genetic engineering.)

In the words of Professor Richard Dawkins, one of the leading New Atheists,

> in a universe of blind physical forces and genetic replication, some people are going to get hurt, other people are going to get lucky, and you won't find any rhyme or reason in it, nor any justice. The universe we observe has precisely the properties we should expect if there is, at bottom, no design, no purpose, no evil and no good, nothing but blind, pitiless indifference.[12]

Science can't answer existential questions; it can't tell us about meaning

Another type of question that science cannot answer includes the more 'existential' questions about ultimate existence and purpose. For example, while science can take us, say, back in time close to the beginnings of the cosmos, it cannot tell us what the first cause of the universe is.

Stephen Hawking, who was perhaps the world's most famous cosmologist, postulated the spontaneous creation of the universe. According to Professor Hawking, the universe began with the Big Bang, which simply followed the inevitable laws of physics. He says:

> Because there is a law like gravity, the universe can and will create itself from nothing … Spontaneous creation is the reason there is something rather than nothing, why the universe exists, why we exist. It is not necessary to invoke God to light the blue touch paper and set the universe going.[13]

This sort of statement is wonderful for newspaper headlines, however, it seems to be a case of passing the explanatory buck from one level of explanation to another. Even if his theory is right, Professor Hawking hasn't explained how the universe comes into existence out of nothing; he has proposed that it comes into existence out of the laws of physics, which existed prior to the universe as we know it. Christians are often criticised by atheists for using God as an ultimate explanation; however, Professor Hawking seems to do the same thing when he uses the laws of physics as if they themselves need no explanation.

Another existential question that faces most human beings at some stage is, "What is the purpose of life?"

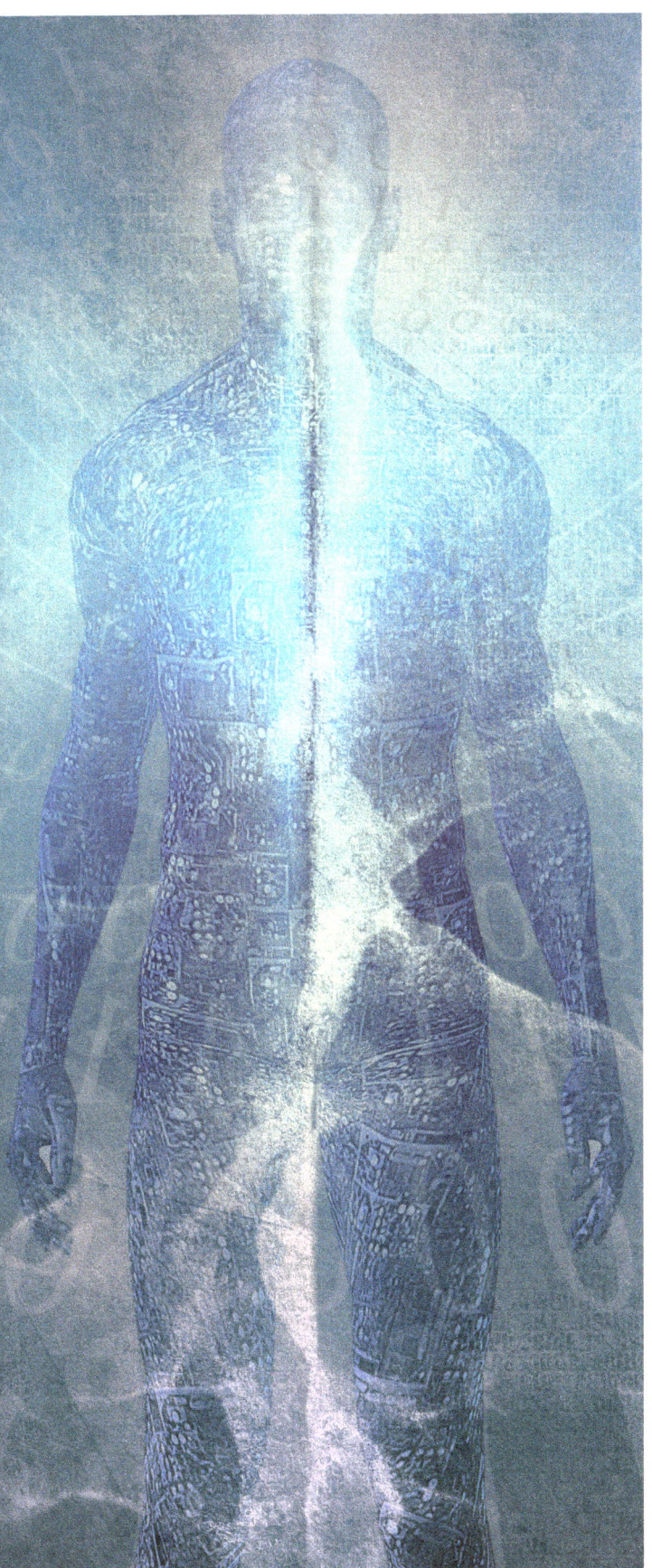

Again, this is not a question that science can answer. Science can tell us, according to the latest understandings of evolutionary theory, how human beings have arisen out of the great chain of biological development over millions of years. Science can tell us that natural selection results in 'the survival of the fittest'. However, even knowing this does not tell us whether there is an ultimate purpose to human life.

A final example of an existential question, which is very pertinent to the discussion of science and faith, is, "Does God exist?" Once again, the reason that this question lies outside of science is because science can only deal with natural processes, causes and objects. However, according to Christians (and other religions) God is not a natural object; God is *super*-natural. There is no scientific observation or experiment that could disprove the existence of God. Lord Martin Rees, astronomer and past president of the British Royal Society (the world's oldest existing society of eminent scientists), puts it this way:

> The preeminent mystery is why anything exists at all. What breathes life into the equations of physics, and actualized them in a real cosmos? Such questions lie beyond science, however: they are the province of philosophers and theologians.[14]

Can science offer certainty?

The lesson of the black swan in Chapter 1 is that (in theory) it only takes one counter-example to overturn a general hypothesis. The implication for scientific theories is that while they can be very well tested, they are open to being challenged by data that contradicts the theory. Hypothetically, Newton's law of gravitation could be 'disproven' if we were only to observe one instance of things 'falling up' – just as one instance of a black swan shows that it is not true that all swans are white. This discussion raises interesting questions about the nature of science: Does this mean there is no final proof in science? Does this mean that all scientific 'truth' could be wrong? Philosophers of science discuss questions such as these.

"In a universe of blind physical forces and genetic replication, some people are going to get hurt, other people are going to get lucky, and you won't find any rhyme or reason in it, nor any justice." Prof. R. Dawkins

Another question raised by the fact that science is dependent on induction (as we saw in Chapter 1) is what is often referred to as 'the problem of induction'. It goes like this: although science must assume that induction is a sound means to arrive at conclusions, we have no way of justifying our confidence in induction. One temptation is to argue that induction is obviously a sound method of arriving at truth because it has worked in the past. However, that would be to fall into a logical trap. That would involve an inductive argument, which relies on 'what has worked in the past' (induction) in order to justify the use of inductive arguments. This is a case of the logical fallacy that philosophers call 'begging the question'. So, it appears that although science depends on induction, it is very difficult to justify this dependence.

What about scientific certainty?

One implication of recognising that science is built on unprovable foundations is that we realise that there is no such thing as absolute proof and certainty. In other words, there is a difference between being able to prove something with certainty and being confident beyond reasonable doubt that something is true. In the law court, the idea of 'proof beyond reasonable doubt' captures the fact that absolute certainty is never possible. The issue of climate change science is a good example: most of the world's climate experts are convinced that human beings are causing significant global warming; however, they would not say that they could prove the case beyond any doubts. In fact, if you insisted on that sort of proof they would probably say that you had misunderstood the nature of science. (See Article #10, "Climate science and Christian belief" on page 33.)

Conclusion

In this chapter, we considered the limits or boundaries of the scientific enterprise. We saw that there are many sorts of questions that lie outside of the jurisdiction of science. While science has proven very powerful in answering questions about the natural world, there are other questions – moral and existential questions, for example – that do not have natural or scientific answers. As well, we saw that science, like all human quests for truth, gives us greater or lesser degrees of confidence about where the truth lies, but it is unrealistic to expect science to offer absolute certainty.

In Chapter 3, we will see what happens if we make the mistake of thinking that all sensible questions can be answered by science. To make this mistake involves changing the way we think about science, turning science into an ideology or an overall belief system. When that occurs, science becomes akin to a religion in which people place their trust for finding the answers to all their questions, including existential and moral ones. This is a view that is often called 'scientism'.

Questions for discussion

Do you think science gives us certainty?

Do you think science can tell us how we should use the discoveries of science and technology?

What is the purpose of life? Can science help us answer that question?

Is there a danger in looking to science for answers to all the questions we might ask?

What similarities do you see between scientific claims and religious beliefs?

The Hubble Space Telescope: Looking back in time

Have you ever looked with wonder at the colourful photos of stars, galaxies and galactic dust clouds such as the Horse Head Nebula in the Orion constellation of our own Milky Way galaxy? Many of these images are produced by the Hubble Space Telescope, named after Edwin Hubble, who showed that the universe is constantly expanding. The Pillars of Creation [pictured] is another Hubble Space Telescope photo that shows a region of the Eagle Nebula, an enormous cloud of galactic dust in which stars are forming.

Telescopes such as Hubble, which is operated by NASA and the European Space Agency, give us a way of looking back in time. This is because the light for Hubble's images has taken many millions of years to reach the telescope. The photos show us what the universe was like in the distant past. The 'oldest' photos, which show us the most distant galaxies, take us back over 13 billion years, close to the time of the beginning of the universe in the Big Bang almost 14 billion years ago.

While science can tell us about the universe's past and (according to our current understanding) that the universe began about 14 billion years ago, it can't tell us why the universe came into existence. No matter how far back science goes, there is always a gap left to be explained. One popular attempt to explain the universe's existence is the book by atheist cosmologist Lawrence Krauss. In his book, *A Universe From Nothing*, Professor Krauss explains how the universe might have come into existence from just the laws of physics. However, even atheist philosophers were quick to point out that Professor Krauss has only pushed the question back. Science is unable to answer, "What caused the laws of physics in the first place?" That is a philosophical or a theological question. According to Christians, such as astronomer, author and speaker, Dr Jennifer Wiseman, (see Article #7 on page 30), it is God who is ultimately responsible for the creation of the universe (including the laws of physics).

> **Science is unable to answer, "What caused the laws of physics in the first place?" That is a philosophical or a theological question.**

'Pillars of Creation', taken by the Hubble Space Telescope

Dr Jennifer Wiseman: Astrophysicist and public speaker

Dr Jennifer Wiseman is an astrophysicist and a Christian who grew up in the country enjoying the natural world, including star-gazing walks with her parents and pets. "The stars were incredible," she says. "We had dark, dark skies there. And I think I was always curious at that point as to what was out there and how could I explore it."[15] The awe Dr Wiseman felt about the universe is still with her as she explores the stars in her job. "There are at least 200 billion stars in our own galaxy, and there are maybe 400 billion galaxies in our observable universe, each one of them with thousands of millions of stars." However, even though astronomers can tell us much about the stars, "comprehending it is something totally different," Dr Wiseman adds.

Dr Wiseman studied astronomy at Massachusetts Institute of Technology and Harvard University. During her studies she discovered a comet, which was named Wiseman-Skiff in her honour. She is currently a senior astrophysicist with NASA studying the regions in our Milky Way galaxy where stars are formed. She is particularly interested in dense interstellar gas clouds (like the Horse Head Nebula pictured below) and the 'cosmic nurseries' where stars are born. She is also active in bringing the wonders and implications of science to the public, and she directs the Dialogue on Science, Ethics, and Religion for the American Association for the Advancement of Science.

Dr Wiseman describes science as a "wonderful tool for understanding the physical universe", but, she says, it is her Christian belief that gives her answers to the bigger questions in life, such as how, in the vastness of the cosmos, humans can be significant at all. For her, human significance "is given as a gift of love from God, who is responsible for the universe". She adds, "I think it's significant that we as human beings can actually investigate the universe, have a sense of our cosmic history, have a sense of our actual connection to the cosmos, and understand it. To me that's a gift." Dr Wiseman is also active in the American Scientific Affiliation, a network of Christians in science, and she frequently gives public talks on the excitement of scientific discovery and the complementarity of science and faith.

> "God is responsible for everything. By studying more of nature you're enriching your understanding of God," she says.

For Dr Wiseman there is no conflict between her science and her faith. In fact, she says, science deepens her faith. "God is responsible for everything. By studying more of nature you're … enriching your understanding of God," she says. This is also the experience of others too: "When I talk to people I find that most people realise that there are deeper questions of life that science can't fully address, and they don't see why there should be any conflict."

And what about the Bible? For Dr Wiseman, the Bible needs to be interpreted because it is not a scientific text: "You have to look at biblical literature from the perspective of when it was written, the original audiences, the original languages, the original purposes and the message that was meant to be conveyed by it."

'Horse Head' Nebula

DNA, the instructions for life

In 1953, scientists Francis Crick and James Watson, partly drawing on the work of others including Rosalind Franklin, proposed the double helix or twisted ladder structure of DNA. Deoxyribonucleic acid (DNA) is the 'master molecule of life'. It carries the coded instructions that control the way our bodies grow from a single cell into the people (or animals or plants) that we are today. In fact, DNA controls the development, functioning and reproduction of every known organism. However, the DNA code does not give us a moral code for how to manage the increasingly complex possibilities that arise from our ability to manipulate the maker's instructions.

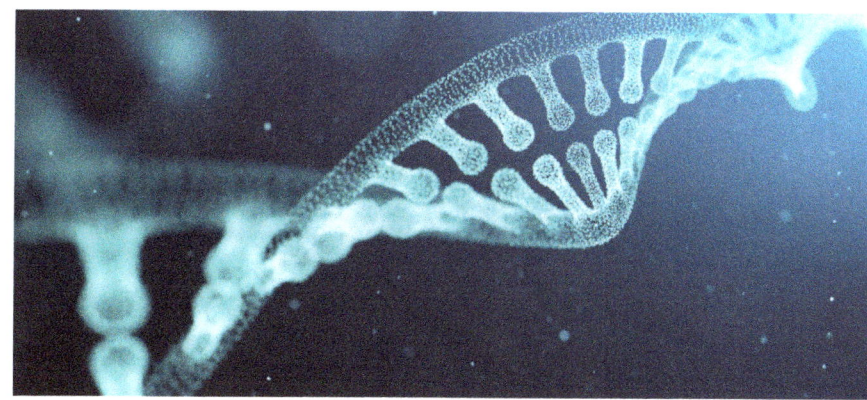

DNA structure

The DNA molecule is like a twisted ladder where every rung on the ladder is made up of two units, called nucleotides, that meet in the middle. There are only four types of nucleotides (A, C, G and T for short), but there are over three billion rungs on a human DNA ladder. In 2003, the Human Genome Project, under the leadership of Francis Collins (see Article #15 on page 58), finished the task of identifying all the nucleotide rungs on the ladder that make up the genetic instruction code for a human being.

Human beings have about 30 million million cells in their bodies (that's 30 trillion), and they are made up of around 200 different cell types. In most of those cells there is about two metres of tightly coiled DNA in the cell's chromosomes. If you add up that two metres of DNA in each cell, your total DNA, uncoiled, would stretch some 60 billion kilometres; that's about 200 times the distance to the sun and back again.

This information, coded into the spiral ladder of the human genome, is inherited from our parents via the single cell (or zygote) that we all started out as. And this DNA template then directs the multiplication and differentiation of cells as we grow, ensuring that you turn into a 'you' and not into a 'me', or into a fish or giraffe. The genetic code determines gender and eye colour and blood type, among other characteristics, and it regulates functions of our body such as digestion. It also plays a part in our abilities, lifespan, height, behaviour … the list goes on.

Since the sequencing of the human genome in 2003, great advances have been made in our ability to manipulate DNA. For example, using a technique called CRISPR/Cas9 we can now cut, splice and add to the DNA of an organism, including humans. CRISPR is very controversial because it has both helpful and worrying possibilities. For example, if a disease is genetically inherited, it might be possible to test embryos and modify their genes so they don't inherit the disease. However, the ability to manipulate the genome could lead to human enhancement that goes beyond simply eradicating disease. Deciding on the line between therapy and enhancement is one difficult question. Another is whether genetic enhancement should be allowed. Should parents have the right to manipulate the intelligence, appearance or the physical ability of their children?

These are questions that science itself cannot answer; they must be answered through discussing moral questions. And for Christians, there are important theological questions involved. For example, if, as Christians believe, there is a purpose to human life, then how can we ensure that the purpose is promoted and not hindered by medical science?

Professor Phil Batterham: Past President of the International Genetics Federation

Professor Phil Batterham runs a genetics research team at the University of Melbourne and he is a past President of the International Genetics Federation. His team is investigating insect genetics and insecticide resistance in order to improve agricultural productivity. Professor Batterham is also a Christian who believes that his science and his faith "fit together pretty neatly".[16] For him, the creation of life by God includes the genetic code, which lies at its core. He sees no tension between the theories of Darwin and the existence of an all-powerful God as described in the Bible.

Professor Batterham affirms that God is the author of all truth. And the key attribute of science is its pursuit of truth about the Earth, life and our universe. Both science and faith are on the same page, he believes, because they both seek to pursue the truth honestly to the best of their ability. He says that the implications of this joint pursuit of truth, for the believing scientist, are that the truths of science should be "incorporated into our theology and they give us an expanded view of God's purpose in this world".

One aspect of genetics that has theological interest is that we now have a genetic window into the past history of the human species. Because we inherit our genes from our parents (and they from theirs, and so on …) genes give a historical record of human lineage on this planet.

We now know that the origins of human beings lie somewhere near the Rift Valley in Kenya. At that time, only a few thousand generations ago, all humans were indigenous Africans. From there, humans started moving out about 60,000 years ago. Australian Indigenous people landed in the north of the country about 50,000 years ago.

> **The key attribute of science is its pursuit of truth about the Earth, life and our universe.**

As a result of this genetic information, Professor Batterham explains, the idea of "human races" based on genetic make-up is "a significant social construct with no biological basis". The DNA code of any two humans on Earth is 99.9% identical. He states that the tiny difference between people's DNA, which makes us all unique, is so small that the idea of different human races is not scientifically justified. Genetically speaking, humans are one race, he says. "We are one large extended family [and] while we are all very similar, we are all unique. God made us to be one family; God made us to be unique. God embraces and loves diversity, so we should too."

Indigenous Australian performer using wooden clapsticks

Oetzi, the Iceman, reconstruction of a prehistoric man

Climate science and Christian belief: A case study

In this article, we will consider the question of climate change for two reasons: it is an example of science at work and it is also an issue that many Christians feel strongly about. This concern for the planet is exhibited by the positive reception of Pope Francis' encyclical *Laudato Si': On Care for Our Common Home*[17], which was published months before the Intergovernmental Panel on Climate Change met in Paris in November 2015.

As an example of science at work, climate change highlights some of the similarities between our scientific convictions and our Christian beliefs. And as an example of a Christian 'issue', caring for the Earth must be based on the best science available, even if it is not absolutely certain.

For Christians, caring for the planet is more than a matter of caring for 'the environment' or ensuring a habitable world for our grandchildren. Christians believe that "the earth is the Lord's and all that is in it" (Psalm 24:1), and so we have a responsibility to look after it. One such Christian is Sir John Houghton, a Welsh Christian climate scientist who was the lead author of the first three reports of the Intergovernmental Panel on Climate Change (IPCC), which is the United Nations body responsible for the scientific assessment of climate change. In 2007, he accepted the Nobel Peace Prize on behalf of the IPCC.

Prior to his work at the IPCC, Professor Houghton was an atmospheric physicist at Oxford University and was head of the British government national weather service (the Meteorological Office). Professor Houghton is also

> **As an example of science at work, climate change highlights some of the similarities between our scientific convictions and our Christian beliefs.**

president of the John Ray Initiative, an organisation that connects the environment, science and Christian belief. Just as he believes scientifically that humans are causing dangerous global warming, Professor Houghton also believes the theological truth that Christians are called to care for the Earth in the way Adam and Eve were called to care for the Garden of Eden (Genesis 1:28).

Professor Mike Hulme is another publicly professing Christian and leading climate scientist. He is the author of *Why We Disagree About Climate Change*. Professor Hulme highlights how difficult climate science and climate policy are because, he says, "all scientific knowledge is partial and provisional".[18] Hulme speaks of "the poverty of scientism" (see Chapter 3) in his field: "the idea that watertight climate science" will be produced that can simply tell us what we should do. Unfortunately, many decisions about climate change need to be made on moral grounds in conditions of uncertainty. "Science alone will not do the policy work that is needed," warns Professor Hulme.

Are there similarities between the way we hold our theological beliefs on the one hand and our scientific beliefs on the other hand? And what is the place of scepticism or belief in climate science and in relation to Christianity?

Often, those who would set up a conflict between science and faith speak as if science is about facts, while religion is about beliefs; science is objective and impersonal, while religion is subjective and personal. However, this sharp distinction between why we believe scientific claims and why we believe religious claims is not as clear as some people suggest. Let's use climate science as an example.

The science of global warming

Scientists are almost all in agreement that global warming is caused by certain gases in the atmosphere, which act as an insulating blanket around the planet trapping heat in a similar manner to the glass walls of a greenhouse. Life on Earth depends on this natural 'greenhouse effect' because without this terrestrial blanket the Earth would be an uninhabitable glacial desert. However, with the industrial revolution of recent centuries we have raised the level of greenhouse gases, particularly carbon dioxide and methane, which increase the effectiveness of the insulating blanket and help to warm the planet.

The issues at the heart of the climate debate are about how significant the human contribution to global warming is and what the consequences are likely to be. These are the key questions tackled by the IPCC, which, since 1988, has collated the research of thousands of scientists under the auspices of the United Nations. According to the IPCC, it is "extremely likely" (95% probability or more) that human beings are causing the significant global warming observed.[19]

In contrast, climate sceptics doubt the IPCC's findings that human activities have a significant influence on global climate. The sceptics would like more evidence and demand 'proof' of human causes.

Now we come to the problem at the heart of climate science, and indeed at the heart of any demand for absolute proof. As Professor Hulme points out, science does not offer that sort of proof. Just as in the case of the IPCC, so all scientific opinion is expressed in degrees of confidence. However, according to scientists such as Professors Hulme and Houghton, in some cases, like climate science, the stakes are enormous, the evidence is beyond reasonable doubt and there is no time for delay.

Climate science and Christian belief

Let's turn now to connections between climate science and Christian belief. On the one hand, theology and science for the most part do not examine the same objects from the same point of view; they pursue truth in different areas and they use different methods to arrive at their conclusions. However, despite these differences, there are many similarities in the way science and theology go about their business.

In the first place, both climate science and Christianity hold their beliefs to be reasonable and true about the things they speak of. While doubts or lack of certainty remain, both science and Christianity make claims that are either true or false. It is either true or false that humans are causing significant changes to the global climate; it is either true or false that Jesus Christ was God incarnate and participated in the creation of the universe. Neither of these claims can be certainly proven to be true or false; both are beliefs held with varying degrees of confidence. "Beyond reasonable doubt" is the term used in the law court.

Second, both scientific and Christian beliefs are based on evidence – neither consists of irrational or blind belief. Science, for the most part, focuses on the empirical evidence of the senses. It weighs up evidence using a number of foundational or 'pre-scientific' assumptions (see Chapter 1). And, wherever possible, science turns to experiment to test its conclusions. On the other hand, Christianity, like historical or moral reasoning, for example, does not rely on repeatable experiments in order to make its case. Rather, just as history and philosophy have their own traditions and norms of enquiry, so too theology is a rigorous discipline with its own methodical use of evidence and reasoning.

Part of the rigour of theology arises from the fact that Christian thought, like climate science, is a collaborative effort based on networks of mutual trust. No one person can master any serious field of enquiry. Thousands of scientists contribute to the climate discussion, each one an expert in their own field. However, they have to trust the judgements of others with whom they collaborate in order to contribute to the overall IPCC 'opinion'. In parallel, no theologian, biblical scholar or philosopher of religion can hope to master any but their own small corner of 'the knowledge of God'.

Within both theology and climate science there is room for further developments and a healthy questioning that challenges the accepted norms in a way that strengthens the edifice of belief – either by confirming it or by showing where some beliefs are found wanting and ought to be rejected. And whether it's science or theology, not only can the evidence be interpreted in different ways (which it often is), but even working out what constitutes significant evidence as opposed to irrelevant 'noise' is an interpretive judgement.

Finally, this room for revision means that neither science nor theology is capable of offering 'final proofs' to the committed sceptic. Renowned scientist turned philosopher, Michael Polanyi, describes scientific belief in a way equally appropriate to religious belief. In disputing 'positivistic' ideas of science and knowledge more generally, Polanyi says science is about achieving "a frame of mind in which I may hold firmly to what I believe to be true, even though I know that it might conceivably be false".[20] As we have seen, this lack of ultimate proof is found in the reports of the IPCC, which speak of degrees of likelihood, reflecting the fact that any scientific statement is, in the final analysis, the considered judgement of fallible human beings. In the words of Nobel Prize-winning physicist Richard Feynman, "scientific knowledge is a body of statements of varying degrees of certainty – some most unsure, some nearly sure, but none absolutely certain".[21]

The climate sceptic, along with many who are sceptical about religious belief, assumes that reasons must be so overwhelming that the believer is left no possibility for doubt. Climate science, like Christianity, can always be questioned further and no amount of evidence will convince the committed non-believer. For most people, however, the end to their scepticism comes when they have heard enough to warrant a conclusion. In this sense, they choose to believe what they also know "might conceivably be false".

In the first place, both climate science and Christianity hold their beliefs to be reasonable and true about the things they speak of. While doubts or lack of certainty remain, both science and Christianity make claims that are either true or false.

CHAPTER THREE

The problem of scientism

In this chapter, we continue exploring the nature of science; in particular, what happens when the valid pursuit of science goes too far and becomes the ideology of 'scientism'. Remember: when we discuss the nature of science, we are not 'doing' science itself; we are thinking about science – about how it works and where it fits into the 'big picture'. The technical name for thinking about science in this way is 'philosophy of science'
(see Article #11, "What is philosophy of science?" on page 45).

One commonly misunderstood issue about the nature of science, which contributes to the myth of conflict between science and religion, is the idea that science is an all-encompassing system that includes all knowledge. This is a view that is often called scientism. In Chapter 1, we considered some of the non-scientific assumptions that science needs to rely on in order to get on with its work. In Chapter 2, we looked at some of the limits of science: questions that science cannot answer (either now, or ever). If science is limited, the following question arises: Why would some people say that all questions and all knowledge should eventually fall under the realm of science?

There are a number of possible answers to this question. Perhaps some people are comfortable with scientific answers, but uncomfortable with 'philosophical', 'religious' or 'supernatural' possibilities. In other words, people might rule out non-scientific answers without thinking about whether that is a rational thing to do. Sometimes this is called 'cognitive bias', which is the idea that our thinking (cognition) is biased in such a way that it only sees certain sorts of answers and so it is biased against considering other possibilities.

Whatever the reasons why people might be confused about how far the reach of science extends, it is helpful to tease out the relationship between science on the one hand, and a 'worldview' called Naturalism, on the other hand.

First of all, let's think about worldviews.

What's in a worldview?

In the hands of some New Atheist thinkers, the naturalistic worldview turns into the ideology of scientism, which argues that science is the only way to truth. Wendell Berry, the Christian farmer and writer, warns of the irony inherent in this sort of thinking: "Whatever proposes to invalidate or abolish religion ... is in fact attempting to put itself in religion's place."[22]

Christianity is a worldview: It's about meanings, not mechanisms

A worldview is a set of ideas and beliefs that offer an overall framework to interpret the universe and the significance of humanity. It's a sketch of the 'big picture'. Imagine that we all 'see' the world through a lens. What

A worldview also helps us answer questions such as: How should we live? What happens after death? Does life have a meaning? Does God exist? What does it mean to be human?

we see will depend on the colour of the lens. If we look at the world through a blue lens then everything takes on a blue tint. So, a worldview 'colours' the way we interpret the world. Here's an example: For a Christian, a human being is intrinsically valuable and made in the image of God; that's the most important way a Christian sees a human person. However, for someone with a Naturalist worldview, the human being might be primarily seen as the most advanced product of evolution. Even if the Naturalist agrees that human beings are 'valuable', this belief will not be based on seeing human beings as being created in God's likeness. In other words, the Naturalist and the Christian interpret human beings (and much else) very differently because of their different worldviews.

A worldview also helps us answer questions such as: How should we live? What happens after death? Does life have a meaning? Does God exist? What does it mean to be human? We could say that a worldview mostly answers questions about meaning. (Remember the boiling water example in Chapter 2, and the difference between meanings and mechanisms?) Although a worldview may not answer every question, it still suggests where the answers might lie and it aims to be coherent in the answers it does give. This means that a worldview cannot contain glaring contradictions within its set of core beliefs.

What sort of beliefs does the Christian worldview consist of? Traditional Christian belief, which is shared by mainstream branches of the Christian church, includes the following:

→ There is a creator God who made the universe and everything in it.

→ Miracles, which we cannot humanly explain, are possible.

→ History is linear from creation through to a final "making all things new" (Revelation 21:5).

→ At the centre of history is the death and resurrection of Jesus Christ, who was both God and man.

→ Humanity, which is shaped in the image of God, has a purpose and is made for a relationship with God.

→ Human beings are not capable of completely comprehending and loving God perfectly, so we are dependent on God both for revelation and for restoring the relationship with our Creator.

We can see from this description of some aspects of the Christian worldview that it mostly addresses questions about meaning and not mechanics; it offers answers to questions about the purposes and not the particles of the universe. It doesn't answer all possible questions, and it would be a mistake to think that it should; just as it would be a mistake to think that science has answers to every type of question.

It would also be wrong to think about Christianity and science as two areas of enquiry that can be placed alongside each other and compared. In the colloquial phrase, that would be like comparing apples and oranges; they are not the same type of thing. A danger of referring to "the science–faith relationship" is that this description seems to set up a symmetry between two comparable entities: science on the one hand and Christianity on the other. However, science and Christianity are not directly comparable because Christianity is a worldview while science, understood rightly, is not and never can be.

Science is not a worldview and Christianity is not science

Science searches for the mechanisms and laws of the universe in the hope of answering the 'how' questions. It looks for the physical causes and constituents of what goes on in our world. However, Christianity is different.

On the one hand, as a worldview, Christianity is much more encompassing than science because it offers answers to the big questions such as: Why are we here? or Why is there something rather than nothing? On the other hand, Christianity has little interest in other sorts of issues such as the 'how' questions. The Bible offers a general foundation for our thinking and acting, but it does not tell us exactly how to run a country or how to order our finances.

Christianity is not science and it is a mistake to think that the Bible is composed to be a political treatise or a scientific textbook. In the words of Galileo Galilei, the central figure in the most famous so-called conflict between science and religion, "The Bible teaches how to go to heaven not how the heavens go."

Science for its part is not a worldview. As we have seen already, physics and chemistry do not make claims about the meaning or purposes of particles or molecules. Biology and astronomy do not tell us the meaning of spiny anteaters or spiral galaxies. That's simply not what they're about, and if we look to science to answer such questions we expect more than it can offer.

Naturalism vs naturalism

Now that we have discussed worldviews and seen that science is not a worldview, we can return to the question of naturalism and why science can be confused for a worldview (scientism). The problem lies in confusing 'methodological naturalism' – which, as we will see, is the basis of science – with a worldview we will call Naturalism (with a capital 'N' to indicate that it is a worldview). Naturalism as a worldview starts from the assumption that the only things that exist are material (materialism) or physical (physicalism).

The windowless room of Naturalism

Naturalism is a worldview that says that the natural world, which science investigates, is all that there is. This view is becoming the dominant way that Western public discourse speaks of reality. According to the Naturalistic worldview, reality is only made up of 'natural' components such as matter and energy. There is no God or gods for Naturalism; there is no non-natural or spiritual realm. As far as human beings and other life is concerned, everything is explained by genes, natural selection and brain chemistry. In its cruder forms Naturalism equates Christianity and other religions to belief in 'fairies at the bottom of the garden' or 'celestial teapots' or the 'Flying Spaghetti Monster'. In other words, for Naturalism, Christians believe in things that don't exist.

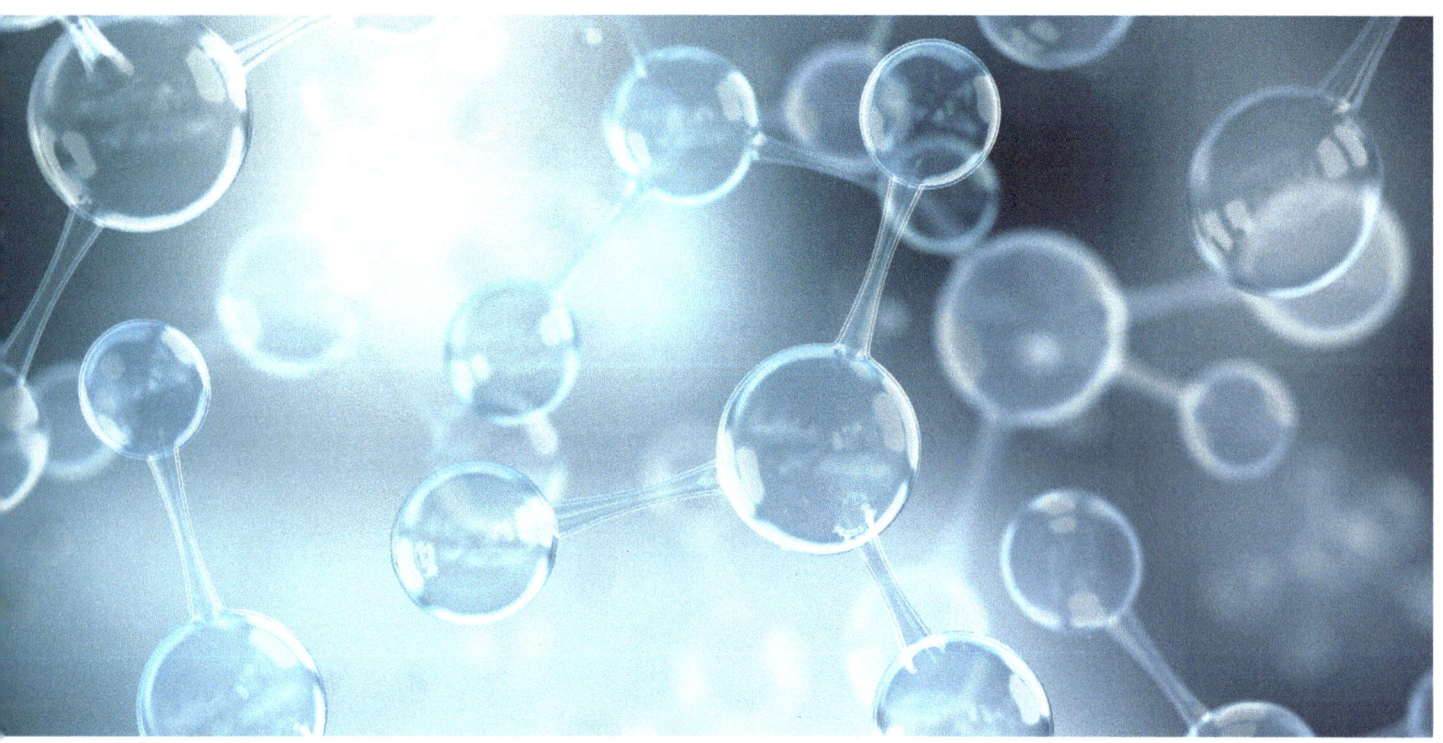

Molecules or atoms

Expressed this way we can see that Naturalism is a worldview competing with other worldviews. It is a belief system that attempts to answer the questions of meaning discussed in Chapter 2, although its answers are of the nihilistic variety, claiming that there is no ultimate meaning. As one author describes Naturalism,

> it's as if we are locked in a windowless room which is brilliantly lit by the scientific method that enables us to see and explain more and more of our physical world but is paradoxically a profoundly reductionist space. It reduces and limits all explanations and descriptions to the material and physical. It has no windows onto wider and bigger explanations of reality. It provides no answers to our deepest and most important questions, like what the meaning and purpose of our lives is, how to understand right and wrong, the nature of justice, beauty, love, shame, guilt, honour, duty, evil and good, why we desire social and personal accountability. The list of enduring human questions it fails to deal with goes on![23]

This 'windowless room' view, which has no space for ultimate morality or purpose, is summed up by the famous atheist Professor Richard Dawkins who describes the universe as having "precisely the properties we should expect if there is, at bottom, no design, no purpose, no evil and no good, nothing but blind, pitiless indifference".[24]

Having now described the Naturalistic worldview, it is important to clarify the difference between Naturalism and the method of investigation known as 'methodological naturalism'.

Science is based on methodological naturalism

At the heart of a good understanding of the relationship between science and Christian belief is the difference between methodological naturalism (small 'n') and Naturalism (the worldview). Methodological naturalism is an essential foundation of science and is definitely not a worldview. It is a tool of the scientific method – a working assumption of science that is used by all scientists whether they are religious or atheist. Methodological naturalism is the assumption that when we do science we are investigating the motion and properties of measurable material phenomena and we are not investigating more profound and elusive questions about the spiritual dimensions of reality. We might say that methodological naturalism assumes that God does not intervene in our experiments.

The role of science is quite appropriately to look for natural explanations and this rigorous pursuit of natural causes and effects was one of the keys to the scientific revolution. This quest for natural explanations means that non-natural causes are ruled out in the laboratory and in scientific thinking. Like a carpenter's hammer, methodological naturalism is an instrument used in order to get on with the job. So, although scientists who use the tool of methodological naturalism may be religious believers, their religious beliefs play no part in the way they do their experiments.

Now that we have clarified what is often a major source of confusion, we are going to do a little more philosophy.

The success of science does not prove that Naturalism is true

Much of the claimed conflict between science and faith arises from confusing the tool of methodological naturalism with a commitment to the worldview of Naturalism. This is particularly evident when people ask a question such as, "But doesn't science disprove religion?"

It seems that what lies behind such thinking is an argument that goes something like this:

- Science is based on naturalism.
- Science is successful.
- So, naturalism must be true.
- But Naturalism and Christianity are contradictory worldviews.
- So, Christianity must be false.

Now, there is a major flaw in this argument. There is sleight of hand where the word 'naturalism' is used in two different ways (note the lower case 'n' and the upper case 'N'). We can see this if we rewrite the first part of the argument more clearly as follows:

- Science is based on 'methodological' naturalism ("God does not intervene in our experiments").
- Science is successful.
- So, 'Naturalism' ("there is no God") must be true.

But as we can see, the conclusion doesn't follow, because the conclusion talks about 'Naturalism' while the first line talks about 'methodological naturalism', which is another thing altogether.

In simple terms, just because science assumes that God does not intervene in experiments (methodological naturalism), does not mean that God does not exist (Naturalism). The success of science can only lead us to conclude this: If God exists, then God normally allows the laws of nature to take their course. So, science seeks truth about the natural world by using the tool of methodological naturalism. Science is not committed to Naturalism, which is a worldview.

Let's turn to the problems of scientism, which is an extreme version of Naturalism and an aberration of true science.

The problems of scientism: Taking science beyond its limits

We suggested in the previous chapter that there are presuppositions of science that underlie scientific practice, but which cannot be arrived at by using science. That is, science cannot show that its own presuppositions are valid. But scientism ignores these subtleties.

Scientism is usually used to describe a naive, almost blind, faith in science. It is the idea that only scientific knowledge is authentic and any other sort of knowledge is meaningless nonsense. It is a 'science has all the answers' or a 'science-only' view of the world and of human knowledge/understanding.

The thinking behind scientism goes something like this: if a Naturalistic worldview is correct – that is, if there is no God or gods and the natural world is all that there is –

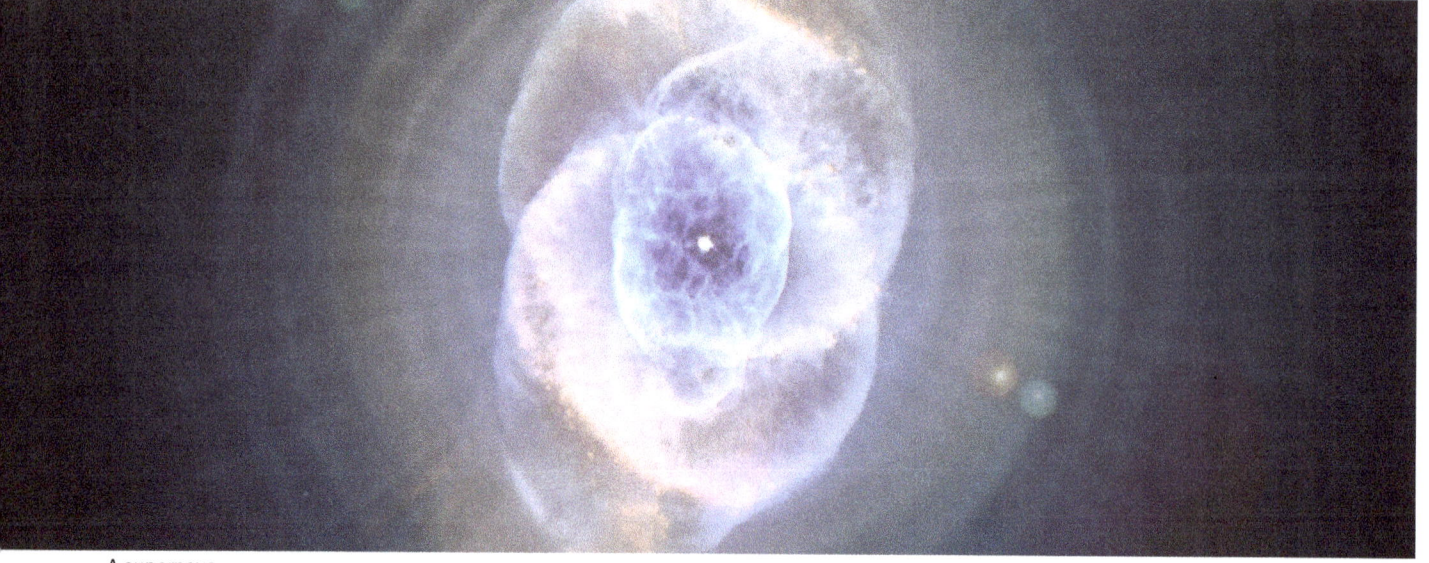

A supernova

Expressed this way we can see that Naturalism is a worldview competing with other worldviews. It is a belief system that attempts to answer the questions of meaning discussed in Chapter 2, although its answers are of the nihilistic variety, claiming that there is no ultimate meaning. As one author describes Naturalism,

> it's as if we are locked in a windowless room which is brilliantly lit by the scientific method that enables us to see and explain more and more of our physical world but is paradoxically a profoundly reductionist space. It reduces and limits all explanations and descriptions to the material and physical. It has no windows onto wider and bigger explanations of reality. It provides no answers to our deepest and most important questions, like what the meaning and purpose of our lives is, how to understand right and wrong, the nature of justice, beauty, love, shame, guilt, honour, duty, evil and good, why we desire social and personal accountability. The list of enduring human questions it fails to deal with goes on![23]

This 'windowless room' view, which has no space for ultimate morality or purpose, is summed up by the famous atheist Professor Richard Dawkins who describes the universe as having "precisely the properties we should expect if there is, at bottom, no design, no purpose, no evil and no good, nothing but blind, pitiless indifference".[24]

Having now described the Naturalistic worldview, it is important to clarify the difference between Naturalism and the method of investigation known as 'methodological naturalism'.

Science is based on methodological naturalism

At the heart of a good understanding of the relationship between science and Christian belief is the difference between methodological naturalism (small 'n') and Naturalism (the worldview). Methodological naturalism is an essential foundation of science and is definitely not a worldview. It is a tool of the scientific method – a working assumption of science that is used by all scientists whether they are religious or atheist. Methodological naturalism is the assumption that when we do science we are investigating the motion and properties of measurable material phenomena and we are not investigating more profound and elusive questions about the spiritual dimensions of reality. We might say that methodological naturalism assumes that God does not intervene in our experiments.

The role of science is quite appropriately to look for natural explanations and this rigorous pursuit of natural causes and effects was one of the keys to the scientific revolution. This quest for natural explanations means that non-natural causes are ruled out in the laboratory and in scientific thinking. Like a carpenter's hammer, methodological naturalism is an instrument used in order to get on with the job. So, although scientists who use the tool of methodological naturalism may be religious believers, their religious beliefs play no part in the way they do their experiments.

Now that we have clarified what is often a major source of confusion, we are going to do a little more philosophy.

The success of science does not prove that Naturalism is true

Much of the claimed conflict between science and faith arises from confusing the tool of methodological naturalism with a commitment to the worldview of Naturalism. This is particularly evident when people ask a question such as, "But doesn't science disprove religion?"

It seems that what lies behind such thinking is an argument that goes something like this:

- Science is based on naturalism.
- Science is successful.
- So, naturalism must be true.
- But Naturalism and Christianity are contradictory worldviews.
- So, Christianity must be false.

Now, there is a major flaw in this argument. There is sleight of hand where the word 'naturalism' is used in two different ways (note the lower case 'n' and the upper case 'N'). We can see this if we rewrite the first part of the argument more clearly as follows:

- Science is based on 'methodological' naturalism ("God does not intervene in our experiments").
- Science is successful.
- So, 'Naturalism' ("there is no God") must be true.

But as we can see, the conclusion doesn't follow, because the conclusion talks about 'Naturalism' while the first line talks about 'methodological naturalism', which is another thing altogether.

In simple terms, just because science assumes that God does not intervene in experiments (methodological naturalism), does not mean that God does not exist (Naturalism). The success of science can only lead us to conclude this: If God exists, then God normally allows the laws of nature to take their course. So, science seeks truth about the natural world by using the tool of methodological naturalism. Science is not committed to Naturalism, which is a worldview.

Let's turn to the problems of scientism, which is an extreme version of Naturalism and an aberration of true science.

The problems of scientism: Taking science beyond its limits

We suggested in the previous chapter that there are presuppositions of science that underlie scientific practice, but which cannot be arrived at by using science. That is, science cannot show that its own presuppositions are valid. But scientism ignores these subtleties.

Scientism is usually used to describe a naive, almost blind, faith in science. It is the idea that only scientific knowledge is authentic and any other sort of knowledge is meaningless nonsense. It is a 'science has all the answers' or a 'science-only' view of the world and of human knowledge/understanding.

The thinking behind scientism goes something like this: if a Naturalistic worldview is correct – that is, if there is no God or gods and the natural world is all that there is –

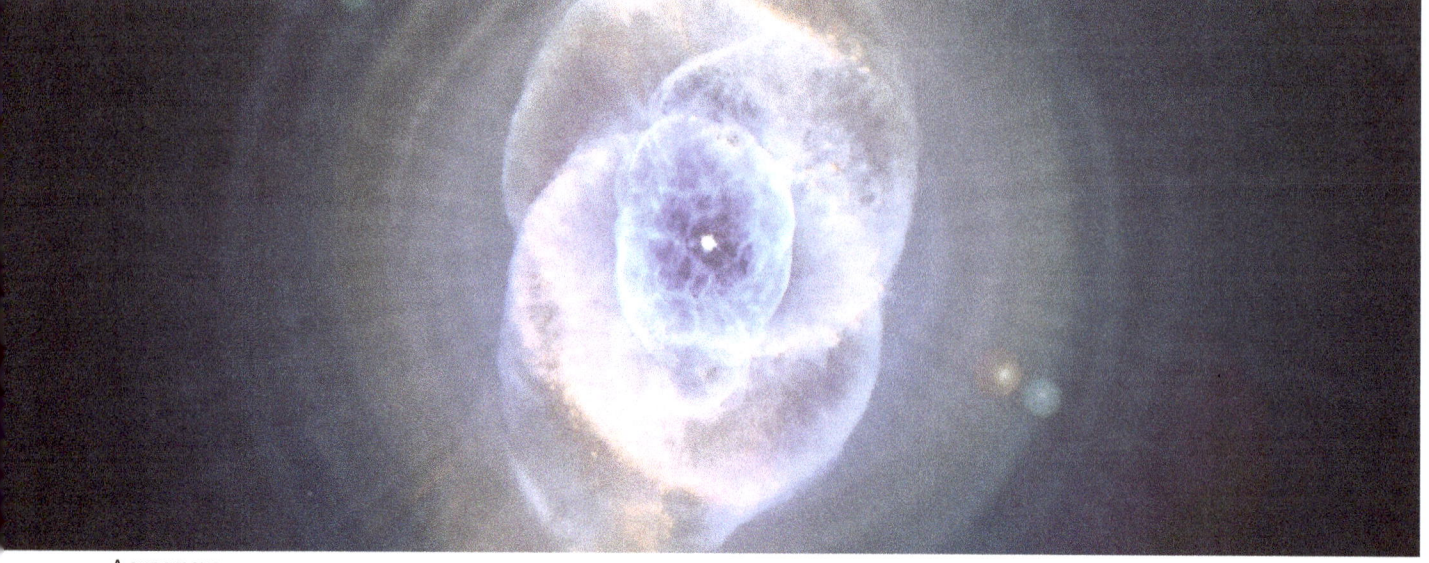

A supernova

(see Chapter 1). And, wherever possible, science turns to experiment to test its conclusions. On the other hand, Christianity, like historical or moral reasoning, for example, does not rely on repeatable experiments in order to make its case. Rather, just as history and philosophy have their own traditions and norms of enquiry, so too theology is a rigorous discipline with its own methodical use of evidence and reasoning.

Part of the rigour of theology arises from the fact that Christian thought, like climate science, is a collaborative effort based on networks of mutual trust. No one person can master any serious field of enquiry. Thousands of scientists contribute to the climate discussion, each one an expert in their own field. However, they have to trust the judgements of others with whom they collaborate in order to contribute to the overall IPCC 'opinion'. In parallel, no theologian, biblical scholar or philosopher of religion can hope to master any but their own small corner of 'the knowledge of God'.

Within both theology and climate science there is room for further developments and a healthy questioning that challenges the accepted norms in a way that strengthens the edifice of belief – either by confirming it or by showing where some beliefs are found wanting and ought to be rejected. And whether it's science or theology, not only can the evidence be interpreted in different ways (which it often is), but even working out what constitutes significant evidence as opposed to irrelevant 'noise' is an interpretive judgement.

Finally, this room for revision means that neither science nor theology is capable of offering 'final proofs' to the committed sceptic. Renowned scientist turned philosopher, Michael Polanyi, describes scientific belief in a way equally appropriate to religious belief. In disputing 'positivistic' ideas of science and knowledge more generally, Polanyi says science is about achieving "a frame of mind in which I may hold firmly to what I believe to be true, even though I know that it might conceivably be false".[20] As we have seen, this lack of ultimate proof is found in the reports of the IPCC, which speak of degrees of likelihood, reflecting the fact that any scientific statement is, in the final analysis, the considered judgement of fallible human beings. In the words of Nobel Prize-winning physicist Richard Feynman, "scientific knowledge is a body of statements of varying degrees of certainty – some most unsure, some nearly sure, but none absolutely certain".[21]

The climate sceptic, along with many who are sceptical about religious belief, assumes that reasons must be so overwhelming that the believer is left no possibility for doubt. Climate science, like Christianity, can always be questioned further and no amount of evidence will convince the committed non-believer. For most people, however, the end to their scepticism comes when they have heard enough to warrant a conclusion. In this sense, they choose to believe what they also know "might conceivably be false".

In the first place, both climate science and Christianity hold their beliefs to be reasonable and true about the things they speak of. While doubts or lack of certainty remain, both science and Christianity make claims that are either true or false.

The open tomb of Jesus, Jerusalem

then the only possible knowledge we can have of anything is scientific knowledge: All that is and all that can be known is verifiable or falsifiable through science; whatever can't in principle be analysed and measured by science is empty belief and fantasy.

In this way, science is held up as the absolute authority in every area of human life and thinking. Instead of science being a tool in the search for truth it has become an ideology – some would say a quasi-religion – that constrains what sort of truths are allowed to exist. New Atheist philosopher Daniel Dennett promotes this view of science when he says, "When it comes to facts, and explanations of facts, science is the only game in town."[25]

Professor Marcelo Gleiser is one well-known scientist who is not a religious believer, but who disagrees with this science-only view of knowledge. He is a theoretical physicist and winner of the US$1.4 million 2019 Templeton Prize for "affirming life's spiritual dimension". Professor Gleiser says that "atheism is inconsistent with the scientific method" because the atheist's hypothesis that there is no God cannot draw on any scientific evidence. "Science can give answers to certain questions, up to a point", so, says Professor Gleiser, we "should have the humility to accept that there's mystery around us". He warns against believing those famous scientists who claim that cosmology has explained the universe and therefore we don't need God: "That's complete nonsense, because we have not explained the origin of the universe at all."[26]

What, then, is the problem with scientism, which says that science can or will have all the answers? To put it bluntly, scientism sounds remarkably like a faith because it boldly claims that science is "the only game in town" or suggests that science can show that Naturalism (the worldview) is correct.

At the heart of scientism lies a logical contradiction. Scientism claims to be rigorously scientific and says that we should believe something like this (let's call it the 'S-thesis'):

The S-thesis: The only claims you should believe are the claims that science shows us to be true.

Now read the S-thesis again. A moment's thought reveals the contradiction of scientism. If we are to believe the S-thesis (that is, that we should only believe scientific claims) then why should we believe the S-thesis itself, which is not a scientific claim? In fact, taking the S-thesis seriously means that we should disbelieve the S-thesis!

In this way scientism seems to be an attempt to lift itself up by the bootstraps, or, to mix the metaphor, the S-thesis shoots itself in the foot. In which case, the appropriate response to Professor Dennett when he says that only science can give us facts, is to ask him simply, "Is that a scientific fact?"

There are many things we believe that are not the result of science. And, as we saw previously, there are many presuppositions that science depends on, but that science cannot show to be true. If science were the only game in town then it wouldn't even get to first base, because the game of science depends on so many 'non-scientific' beliefs.

However, as most scientists, religious or otherwise, know well, science is not scientism and scientism does not follow from science: it is one thing to affirm the validity of scientific knowledge, but it is another thing to say that all knowledge must be scientific.

What about miracles? Did Jesus rise from the dead?

In our discussion above it is clear that if you are committed to scientism – that is, the view that everything has a scientific explanation – then you will have difficulty believing in miracles. This is because miracles are normally understood as events that cannot be explained by the natural laws of the universe.

According to Christians, the most famous and significant miracle in history was the resurrection of Jesus on the third day after he was crucified and died. Science has no explanation for this type of event, so, if you have a presupposition that *every* event has to have a scientific explanation – the view we called scientism – then you would say that this miracle did not occur: either Jesus did not rise or he was never dead.

However, if you think that there might be explanations lying outside scientific explanations, then you would acknowledge at least the possibility of miracles, even though they would not be explainable or verifiable scientifically. The question of whether miracles are possible is not a scientific one; it is a question concerning what presuppositions you have about whether science "is the only game in town". In the words of Francis Collins (see Article #15 on page 58), "a discussion about the miraculous quickly devolves to an argument about whether or not one is willing to consider any possibility whatsoever of the supernatural".[27]

Conclusion

In this chapter, we have considered the attempt to turn science into an all-encompassing worldview known as scientism. We saw that this is an illegitimate extension of the boundaries of science. We also saw that there is a logical contradiction in claiming that all truth must be scientific – it's a bit like being confined to a windowless room and, because you can't see what is outside the room, saying that there is nothing there.

In the next chapter we will look at the prevalent idea that there is a fundamental conflict between science and Christian belief. We will see that this 'conflict thesis' is based on historical misunderstandings, as well as on the misunderstandings that we considered in this chapter and in Chapter 2 about the nature of science itself.

Jesus and Mary Magdalene, 'Noli Me Tangere', Antonio da Correggio

Questions for discussion

Can you think of an example of cognitive bias, where you might tend to misinterpret evidence in a way that confirms your own beliefs? (One example might be the way you interpret a brief email or text message if you already think the person is annoyed with you.)

What is a worldview?

Describe some of the elements of your own worldview.

How would you describe scientism to someone who had never heard the word?

What do you think science can tell us about the possibility of miracles?

What is the difference between Naturalism (the worldview) and methodological naturalism?

What is philosophy of science?

If science is the study of nature, philosophy of science is the study of science. It is the area of study that asks about what counts as science, how science works, what the foundations of science are and what we can hope for from scientific endeavour. Doing philosophy of science is not the same as doing science. Someone can be an expert in their field without being an expert about their field. To give an example, Usain Bolt may be an expert runner, but he is not an expert at the physiology of running. A great flautist has mastered playing the flute, but they may not have mastered the physics of soundwaves. So, too, someone like Professor Lawrence Krauss (see Article #12, "A New Atheist example" on page 52), may be a very good scientist but not have a very good grasp on the philosophy of science.

Some of the questions that philosophers of science probe include the following:

What is science? What is the boundary between science and non-science or pseudo-science? For example, is psychology science? Is economics? What about astrology? Why? Why not?

> **Doing philosophy of science is not the same as doing science. Someone can be an expert in their field without being an expert about their field.**

What does it mean to say something is a scientific law or a law of nature? Can we use the word 'law' beyond physics and chemistry; for example, in areas such as biology and psychology?

Is there a common method to the sciences?

How do we identify the difference between cause and coincidence? How do we know when one type of event (like the Moon's position relative to the Earth) causes another (like the tides), or whether it is simply a case of one event happening after another?

How do scientists arrive at theories or hypotheses? Are there rules for doing so? Or do they result from an uncontrolled flash of inspiration?

What sort of evidence is enough to confirm a theory?

Is it valid to rely on induction? (See the discussion of induction in Chapter 1.)

How much trust should we put in scientific models that we know are incomplete approximations? For example, should we trust climate models that predict global temperatures for the year 2100?

Moon phase dial, St Mark's Clocktower, Venice

A supermoon

CHAPTER FOUR

"I believe in science so ..." The myth of conflict

"I believe in science so I couldn't be a Christian" – that's the phrase heard so often when the question of science and Christian belief arises. By drawing on the previous discussion of the limits of science, this chapter aims to tease out why 'believing in science' does not lead to rejecting a Christian worldview.

So far we have seen that the Bible and the Christian worldview offer answers to what we might call questions of meaning and purpose. For its part, science offers insight into the wonders of the natural world, as well as greatly benefiting the common good. We have also seen that both science and Christian theology have their limits: science can't answer moral, philosophical or existential questions, and, as St Augustine warned, the Bible doesn't answer scientific questions. With this understanding of the nature of science and of Christianity, we see that science and Christian belief can be compatible and even complementary, and historically they have got on quite well. Having laid this groundwork we turn to tackle the so-called conflict thesis head on. Why is it so common to hear that we have to make a choice between taking science seriously and being a serious Christian?

In this chapter, we will consider the current cultural climate in the Westernised world, especially the strand of atheism known as 'New Atheism', and then we will go back to historical and philosophical roots. We will confirm what we saw in the last chapter: it is the ideology or worldview of scientism, not science itself, that is in conflict with Christian belief.

The sources and celebrities of the New Atheism

The so-called New Atheism is actually an old idea rebadged for a digitally connected world – a world where the popularity of celebrity speakers often far outstrips the soundness of their ideas. We have already mentioned some of the best known of the New Atheists, biologist Professor Richard Dawkins and cosmologist Professor Lawrence Krauss. Both have written popular books promoting atheism and both are well known online and on the global lecture circuit.

A clarification: we should note that it is important when discussing the New Atheism to be clear that we are not discussing all atheist thinking. In fact, there are atheists who are concerned about the shallowness of New Atheist thinking. The highly respected philosopher of science and atheist Professor Michael Ruse says the New Atheism is damaging to the cause of atheism and, in reference to Professor Dawkins' book, Ruse says "*The God Delusion* makes me embarrassed to be an atheist."[28]

Ideologically speaking, the sources of the New Atheism lie in what came to be known as the positivist philosophy of the early twentieth century. The positivists or logical empiricists followed the philosopher David Hume's advice when he urged that only mathematics, logical reasoning and the evidence of the senses should be accepted, and all other pursuits are nonsense. Hume's view is famous:

> If we take in our hand any volume; of divinity or school metaphysics, for instance; let us ask, Does it contain any abstract reasoning concerning quantity or number? No. Does it contain any experimental reasoning concerning matter of fact and existence? No. Commit it then to the flames: for it can contain nothing but sophistry and illusion.[29]

For New Atheists such as Professor Krauss, all religious belief is akin to believing in the tooth fairy or Bertrand Russell's 'celestial teapot'. Professor Krauss thinks that it is important to "encourage people to replace the kind of things they get from religion with things that are related to the real world and not myths and fairy tales".[30] (See Article #12, "A New Atheist example" on page 52)

Rumours of divorce: Understanding the conflict thesis

Let's more closely examine the idea that science and faith are at war. This theory, known as the 'conflict thesis' is relatively new historically and dates to the mid-nineteenth century when discussions specifically about 'science and religion' and their possible conflict started to occur. Prior to this, the word 'science' included much that we do not think of as science today, including some aspects that would now fall under the realms of philosophy or religion.

In 1874, the scientist John William Draper published a book called *History of the Conflict between Religion and Science*[31] and, in 1896, Andrew Dickson White, a historian, published *A History of the Warfare of Science with Theology in Christendom*.[32] Henceforth, the conflict thesis, so named after Draper's book title, is often referred to as

---

The so-called New Atheism is actually an old idea rebadged for a digitally connected world – a world where the popularity of celebrity speakers often far outstrips the soundness of their ideas.

Christ mosaic on the façade of the Duomo Santa Maria Del Fiore, Florence

the Draper–White thesis or, based on White's title, the warfare thesis. The preface to Draper's book sums up the thesis:

> The history of Science is not a mere record of isolated discoveries; it is a narrative of the conflict of two contending powers, the expansive force of the human intellect on one side, and the compression arising from traditionary faith and human interests on the other.

It is true to say that there have always been skirmishes that could be described as being between science and religion. And, historically, there is no shortage of people who claim that science and religion are incompatible. However, as the examples in this book confirm, there is also no shortage of prominent religious and non-religious scientists, as well as atheist philosophers of science and Christian scholars, who affirm that there is no necessary or insurmountable conflict.

Perhaps the language of marriage and divorce will clarify the question. In most marriages some difference of opinion is bound to occur, but that does not mean that divorce is imminent. A little conflict is a normal part of a relationship, and it would be wrong to say that historical examples of conflict mean that the relationship is impossible to maintain.

Let's talk about two different understandings of the conflict thesis, which we will call the 'historical skirmish thesis' and the 'philosophical/theological conflict thesis'.

The historical skirmish thesis is the claim that there have been skirmishes in the past that we might describe (in retrospect) as being conflicts between science and religion. This is not controversial, although the examples do need to be examined carefully. As well, we should remember that even using the words 'science' and 'religion' in this way is a historically new way of describing things: those are relatively new categories and new ways of dividing up human intellectual endeavour.

The two crucial and most famous historical examples that are cited in favour of the conflict thesis are the 1860 Oxford evolution debate (see Article #13, "Evolution and the Bible" on page 55) and the Galileo affair (see Article #14 on page 56) both of which are more complicated than the way in which they are described by the proponents of the conflict thesis. However, whatever examples of historical skirmishes that might be found, they only point to specific examples of disagreements between people and groups, and they do not necessarily deal with the more substantial question: Is there something *intrinsically* in conflict between science and Christian belief?

The philosophical/theological conflict thesis is the claim that there is something about the nature of science and religion that make them fundamentally incompatible. This view goes beyond historical examples of disagreement and addresses the philosophical and theological arguments for an insurmountable conflict. Such arguments come from two sides: atheists and theists (usually Christians).

In the last chapter, we examined the misunderstandings about the nature of science that lead (wrongly) to the idea that there is a conflict with Christianity. Article #12 on page 52 of this current chapter continues the description of the problems of scientism by looking at the example of Professor Lawrence Krauss. In that article we see that while Professor Krauss might be an excellent cosmologist, he goes beyond his field of expertise when he makes claims about religion and philosophy.

The atheist arguments in favour of a conflict thesis resemble a type of atheistic 'faith'. We might describe such views as a faith because they rest on a commitment to beliefs that go beyond what science is able to show (see those we considered in Chapter 1 and in the article about Professor Krauss). Such beliefs are based on presuppositions that are held by faith; for example, that natural science and empirical methods are the only trustworthy sources of knowledge.

The Christian version of the conflict thesis

There is also a particular version of the conflict thesis promoted by some Christians who believe that there is a fundamental conflict between the truth revealed in the Bible and the knowledge claims of science, particularly the theory of evolution. This minority view among Christians has arisen particularly in the last 100 years and is often referred to as Young Earth Creationism (YEC). Sometimes YEC is simply called 'creationism', but this is misleading because all Christians believe in God as Creator.

The historically accepted and majority view among Christians is often referred to as Theistic Evolution, pointing to the fact that behind all creation, including biological evolution, is God the Creator. Pope Francis affirms this view putting it this way:

> When we read the account of Creation in Genesis we risk imagining that God was a magician, complete with an all-powerful magic wand. But that was not so. He created beings and he let them develop according to the internal laws with which he endowed each one … And thus Creation has been progressing for centuries and centuries, millennia and millennia, until becoming as we know it today, precisely because God is not a demiurge or a magician, but the Creator who gives life to all beings … The Big Bang theory, which is proposed today as the origin of the world, does not contradict the intervention of a divine creator but depends on it. Evolution in nature does not conflict with the notion of Creation, because evolution presupposes the creation of beings who evolve.[33]

However, YEC holds the position that scientific views of evolution and the age of the earth are wrong. YEC is based on a very particular way of reading the Bible that might be described as literalistic because it does not recognise the diversity of text type found in the Bible (see the discussion of genre in Article #2 on page 6). If we try to read the creation accounts in the first chapters of Genesis as science or as a source of facts about the natural world, we will definitely find conflict with scientific

Sculpture of Galileo Galilei, Aristodemo Costoli, Florence

knowledge. This way of approaching the Bible does not pay due respect to the intentions of the biblical authors (who were not scientists teaching physics or geography), and it does not recognise that the authors naturally used the language and conceptions of their own day.

Conclusion

This chapter briefly considered aspects of the sorts of conflicts between science and Christian belief that people have proposed. We saw that although there have been historical conflicts, like the Galileo controversy, they are probably better described as skirmishes because they do not raise insurmountable problems between faith and science.

We also considered the possibility of other, more fundamental sorts of conflict that might be described as philosophical or theological. In this chapter and the previous one we considered some of the misunderstandings about the nature of science that might give rise to the conflict thesis. The article on the New Atheism and Professor Lawrence Krauss teases out this issue in more detail.

Finally, we recognised that some Christians interpret the Bible, particularly the first chapters of Genesis, in a literalistic fashion that conflicts with mainstream science. This YEC is not the commonly held view among Christian theologians or scientists who are Christians, nor is it the historically mainstream view. We saw how this position was partly based on a misunderstanding about the genre or text type of certain parts of the Bible.

Questions for discussion

Would you describe the relationship between science and Christianity as harmonious or are there irreconcilable differences? Why?

What is the difference between historical conflict and a philosophical or theological conflict?

Before reading this book, did you agree with the conflict thesis? Did you assume there was a fundamental conflict between science and religious faith?

How has your view about the conflict thesis changed?

A New Atheist example: Lawrence Krauss and *A Universe from Nothing*

One of the most prominent of the New Atheists is cosmologist Professor Lawrence Krauss, the author of *A Universe from Nothing: Why There Is Something Rather than Nothing*.[34] As an example of strident atheism, we will look briefly at some of his thinking, noting how he strays from his field of expertise in science to philosophy and theology.

In *A Universe from Nothing*, Professor Krauss claims that the universe could have brought itself into existence out of nothing – that is to say, out of just the laws of physics. One implication of this, he says, is that there is no need for a divine creator to explain the existence of the universe. When the book was published, philosophers and theologians were quick to point out the problem with this argument: the laws of physics are something and not 'nothing'. What Professor Krauss has done is to merely push the question of the origin of the universe one step further back so that the question of origins becomes "Why do the laws of physics exist rather than nothing?" For Christians, God has always existed and needs no cause – unlike the entities that science investigates. So, Christians believe that God is the 'first cause' who is responsible for the laws of physics.

Professor Krauss has been pressed about his definition of 'nothing' by critics who suggest that he is redefining the word to mean something physically or materially present. His answer is that "'nothing' is a physical concept because it's the absence of something, and something is a physical concept". Is he right? Normally we think that there are lots of 'somethings' in the world that are not physical somethings. Someone's love for their spouse, the number 42, and God (Christians believe) are all something rather than nothing, but none of them are physical concepts.

The important thing to notice in this argument about 'nothing' is

> **Normally we think that there are lots of 'somethings' in the world that are not physical somethings.**

that Professor Krauss moves from his scientific field of expertise to philosophy of science when he claims that science can offer answers to questions that are philosophical or existential. For him, religion is nonsense: there is a fundamental conflict between science and faith, and the only way he can see the world is through the lens of science. He says:

> Science is incompatible with the world's major religions; all of those are, from a scientific perspective, nonsense. God is irrelevant. People seem to think it's an important question; it's not an important question to scientists. God isn't necessary to discuss the universe.[35]

As we have seen, it is true in one sense to say that God is not necessary to investigate the natural world. Atheists and religious believers do science side by side. However, Professor Krauss makes a stronger claim than that. Despite his words quoted above, he believes in the 'science is everything' view that we have called scientism. For him, science answers all the questions because he does not recognise any boundary between science and other areas of human thinking. For him, not only does science provide understanding of the physical world, but it also offers its own spirituality and is the foundation of morality. He states:

> Scientific empiricism – rational thought combined with empirical enquiry, which is the way we learn about the world – brings much more meaning and spiritual wonder than religion ever does.
>
> … Science can enhance your appreciation of your place in the cosmos and of course provide a much sounder moral framework – a framework for determining what's right and wrong – than religion.

Empirical evidence and the naturalistic fallacy

One of the errors that Professor Krauss makes is an example of what philosophers call the 'naturalistic fallacy' (often stated as "you can't get an *ought* from an *is*"), which is the attempt to draw moral conclusions from the facts of nature based on the idea that science can give us evidence for moral truths. For example, if science shows that biological evolution is a savage survival of the fittest does that mean we should draw our morality from these biological facts? Jesus' commandment "you shall love your neighbour as yourself" (Mark 12:31) clearly stands against any moral norms that might be drawn from biological evolution.

"You can't get an *ought* from an *is*."

The thinking that lies behind Professor Krauss's thinking has to do with his understanding of the type of evidence we need to believe something. When asked about why he believes only the things that science reveals he shows a very limited understanding of what might count as evidence. He states:

> If there is empirical evidence for something then I'm willing to accept that fact. Nothing is going to unsettle that because that works. If I looked up tonight and the stars rearrange themselves to

say "I am here" then it would be worth thinking about. But as there has been no single shred of empirical evidence for any deity or any divine intervention in the history of the universe, there's no reason to worry about it.

So Professor Krauss relies only on empirical evidence, usually understood as the evidence of the senses: what we can see, hear, taste, touch and smell. In doing so, he casts all religion and all moral discussion into the flames because for him it is meaningless. He also casts aside the historically attested evidence of Jesus' life, death and resurrection. After all, there is no empirical evidence of the sort Professor Krauss seeks for the existence of any historical figure. Nor is there that sort of evidence for the equality of human beings, the wrongness of torture or the meaning of life. As well, there is no empirical evidence for the presuppositions that science is based on (discussed in Chapter 1), as the movie *The Matrix* highlights; we could (hypothetically) be part of the matrix that manipulates our five senses to make us think we live in the real world.

Another challenge for Professor Krauss's view is that much of science, including especially his own field of physics, depends on constructing theoretical models, as well as on empirical data. However, the judgement that one model or theory is better than another is based on much more than what the empirical data reveals. In science, choosing the best theory is based on less tangible criteria such as the beauty or simplicity of theory, or whether it is internally coherent, and on whether it has the power to explain the data better than another theory.

Another challenge for Professor Krauss's view is that much of science, including especially his own field of physics, depends on constructing theoretical models, as well as on empirical data.

For New Atheists like Professor Krauss, science is no longer a worthy vocation for studying nature within a bigger worldview that contributes meaning and purpose. It becomes the windowless room of scientism, a worldview that proclaims that all that is not empirical is nonsensical.

When asked about respected scientists who are also serious Christians, Professor Krauss suggests that scientists who are religious are refusing to deal with the facts. "You can be a scientist and a Christian, but to do that you have to suspend your disbelief," he says. This view echoes Professor Dawkins' opinion that "religion poisons your ability to use your brain".[36]

Perhaps it is Professor Krauss who is indulging in narrow thinking? In an interview he was asked if there was a danger, for him as a physicist, of wading into the depths of philosophy and theology. He answered:

No. Because there are no depths of philosophy and theology. I don't wade into philosophy and theology; I don't need to. You don't need to know anything about philosophy or theology to do physics. Or to understand the universe.

What Professor Krauss doesn't seem to see is that he makes many philosophical and theological statements and that his popularity as a speaker is because he mixes science with philosophical and theological pronouncements. In fact, his response that "you don't have to know anything about philosophy … to do physics" is not an empirical statement. It is not a scientific statement, but one that belongs in the philosophy of science.

Prof. Lawrence Krauss

Evolution and the Bible

Apart from the Galileo affair (See Article #14 about the Galileo affair on page 56), the most famous example of conflict between science and faith stems from the 1860 Oxford evolution debate, which occurred just seven months after Charles Darwin published perhaps the most famous scientific book of all time, *On the Origin of Species*. And, like the Galileo story, this one too is often painted as if it were a straightforward battle between the advance of science and the resistance of religion. In fact, the Church of England at the time was supportive of science and many English scientists were religious figures who used theology to popularise science.

In 1859, Darwin's *Origin* had proposed a mechanism that might allow a species to evolve into another species. That mechanism was natural selection, mentioned in the full title of Darwin's work: *On the Origin of Species by Means of Natural Selection, or the Preservation of Favoured Races in the Struggle for Life*. However, the idea of one species evolving into another one met significant resistance in both scientific and church circles.

Although the debate at the Oxford University Museum of Natural History involved many speakers, it is also called the Huxley–Wilberforce debate after the scientist Thomas Henry Huxley and the Anglican (Church of England) Bishop Samuel Wilberforce.

Huxley, also known as 'Darwin's bulldog', was a strong supporter of Darwin's views, partly for social and cultural reasons. Huxley wasn't one of the elite, educated at Oxford or Cambridge universities. At the time, to go to those universities, you had to sign up to Anglicanism. Huxley resented the Church of England's influence over science in Britain, so he was keen to separate science from religious control. To undermine the authority of the Church and set up a church–science conflict, Huxley explicitly used evolutionary theory.

Huxley's attempts succeeded to some extent; however, Darwin's views also had significant Church support from the establishment. So, once again, we see that the simple story of science versus religion misrepresents the actual complex situation.

Today, almost 200 years after the famous Oxford evolution debate, Dr Francis Collins, as an eminent scientist committed to biological evolution, sums up one Christian understanding of evolution as part of a bigger religious worldview:

"There are many subtle variants of theistic evolution, but a typical version rests upon the following premises:

1. The universe came into being out of nothingness, approximately 14 billion years ago.

2. Despite massive improbabilities, the properties of the universe appear to have been precisely tuned for life.

3. While the precise mechanism of the origin of life on earth remains unknown, once life arose, the process of evolution and natural selection permitted the development of biological diversity and complexity over very long periods of time…

4. Once evolution got under way, no special supernatural intervention was required.

5. Humans are part of this process, sharing a common ancestor with the great apes.

6. But humans are also unique in ways that defy evolutionary explanation and point to our spiritual nature. This includes the existence of the Moral Law (the knowledge of right and wrong) and the search for God that characterizes all human cultures throughout history."[37]

Clearing the air about the Galileo affair

Perhaps the most famous example of so-called conflict between science and faith is the 'Galileo affair'. It involved a series of events in the early 1600s that resulted in the scientist Galileo Galilei being tried in 1633 by the Roman Catholic Church. In fact, the story is complicated and often portrayed in order to promote the idea of a faith and science conflict; however, Galileo was a faithful Catholic and the Roman Catholic Church had good science on its side. There are various interpretations of the Galileo affair: those opposing the Catholic Church are very critical of the Church, while those supporting the Church point out the many ways the Church was trying to resolve the situation.

Professor Peter Harrison is an Australian academic and Christian who directs the Institute for Advanced Studies in the Humanities at the University of Queensland. He is globally recognised as an expert on the scientific revolution, including the relationship between science and religion. Professor Harrison says that the Galileo affair is "not a straight-forward conflict between science and religion" and that "in no way was this typical of the Catholic attitude to science."[38]

Professor Harrison highlights that although Galileo was right in challenging the Earth-centred view of the universe, he was going against the best science of the time: "the scientific consensus at the time was firmly against the view that Galileo was proposing". Actually, "The church did a lot of work to try to get the science right and to that extent they were essentially supporting

> **Perhaps the most famous example of so-called conflict between science and faith is the 'Galileo affair'.**

what they believed to be the correct scientific consensus … there are a number of difficulties with the scientific claims that Galileo was making." For these scientific reasons, "the Catholic Church and indeed the scientific community more broadly thought

'Galileo facing the Roman Inquisition', Cristiano Banti

The book, *Dialogue Concerning the Two Chief World Systems*, written by Galileo in 1632, comparing the Copernican system with the Ptolemaic system

that the hypothesis of a moving earth was deeply problematic". In fact, says Professor Harrison,

> from the late Middle Ages and probably into the 18th century, no other institution in Europe had supported astronomical research to the extent that the Catholic Church did. They were big supporters of astronomy and obviously the medieval universities which the Catholic Church got up and running; this was the site where science was conducted. They were huge supporters of knowledge and learning.

As well as the scientific matters at stake, another factor in Galileo's conflict with his church was that he used Scripture to support his position. "This was a disastrous move politically," Harrison believes, because, for individuals to claim the right to interpret Scripture was a Protestant position; it had been explicitly condemned at the Council of Trent and the Catholic Church was very nervous about people essentially going down this Protestant path. So, that's part of the background.

Meanwhile, there was also a background of politics within the Church with some supporting Galileo and others against him. The Pope himself, Urban VIII, was originally a good friend and supporter of Galileo until, in one of his books *(Dialogue Concerning the Two Chief World Systems)*, Galileo appeared to make the Pope appear foolish by putting his words in the mouth of a character called Simplicio (which in Italian can mean simpleton or fool). In the end, Galileo, the difficult personality, who promoted a view against the science of the time, and who had both friends and enemies in the Church, was placed under (rather comfortable) house arrest. And, Harrison emphasises, "he remained a committed Catholic to the end of his life; in fact, Galileo was a member of the clergy".

Pope Urban VIII, Pietro da Cortona

Dr Francis Collins: Medical researcher and genetic pioneer

Dr Francis Collins is a medical doctor who was the director of the Human Genome Project and, following that, the director of the US government National Institutes of Health (NIH). The Human Genome Project was the 15-year effort to identify the three billion 'letters' of the human DNA instruction book (see Article #8, "DNA, the instructions for life" on page 31). After the project was completed, Dr Collins became head of the NIH, which is the largest medical research agency in the world. He was awarded the US Presidential Medal of Freedom in November 2007 and the National Medal of Science in 2009. Dr Collins is also a well-known author for his bestselling book *The Language of God: A Scientist Presents Evidence for Belief*, in which he offers various arguments in favour of the existence of God and in particular shows how biological evolution fits with his Christian belief.

According to Dr Collins, science is a worthy pursuit summed up as "mankind trying to understand the greatness of God's design". In fact, he says, if we do not respect science we do a disservice to God because

> the God of the Bible is also the God of the genome. He can be worshipped in the cathedral or in the laboratory. His creation is majestic, awesome, intricate and beautiful, and it cannot be at war with itself. Only we imperfect humans can start such battles. And only we can end them.[39]

For Dr Collins, the experience of "sequencing the human genome, and uncovering this most remarkable of all texts, was both a stunning scientific achievement and an occasion of worship".[40]

As a scientist who thinks that humans have come about through the evolutionary process of mutation and natural selection, Dr Collins asks if that means we have no need for God to explain humanity. His own reply raises the question of what it means to be human:

> The comparison of chimp and human [DNA] sequences, interesting as it is, does not tell us what it means to be human. In my view, DNA sequence alone … will never explain certain special human attributes, such as the knowledge of the Moral Law and the universal search for God. Freeing God from the burden of special acts of creation does not remove him as the source of the things that make humanity special, and of the universe itself. It merely shows us something of how he operates.[41]

Dr Collins states that faith and reason are not and never have been mutually exclusive. He asks, "In this modern era of cosmology, evolution, and the human genome, is there still the possibility of a richly satisfying harmony between the scientific and spiritual worldviews?" He answers his own question "with a resounding yes":

> In my view, there is no conflict in being a rigorous scientist and a person who believes in a God who takes a personal interest in each one of us. Science's domain is to explore nature. God's domain is in the spiritual world, a realm not possible to explore with the tools and language of science. It must be examined with the heart, the mind, and the soul – and the mind must find a way to embrace both realms … I was vaguely aware that some of those around me thought that this pairing of explorations was contradictory and I was headed over a cliff, but I found it difficult to imagine that there could be a real conflict between scientific truth and spiritual truth. Truth is truth. Truth cannot disprove truth … I do not believe that the God who created all the universe, and who communes with His people through prayer and spiritual insight, would expect us to deny the obvious truths of the natural world that science has revealed to us.[42]

The Roman Catholic Church and science

The Roman Catholic Church has dealt with science and faith in a systematic way for centuries, and it has been a driving force in promoting science through that time. The Church has founded many universities, schools and hospitals, and many Roman Catholic clergy have also been scientists. They include Gregor Mendel, the monk who founded the field of genetics, and Georges Lemaître (see Article #4, "Georges Lemaître: Priest and father of the Big Bang" on page 20). Today, the Vatican Observatory is an astronomical research and educational institution, and a leading contributor to science.

In 1893, Pope Leo XIII wrote in an encyclical that "no real disagreement can exist between the theologian and the scientist provided each keeps within his own limits",[43] and he affirmed St Augustine's warnings and advice of some 1500 years earlier (see Article #1, "The Bible is not science: St Augustine's warning about confusing science and faith" on page 5).

The Catechism of the Catholic Church, which is an official summary of Roman Catholic doctrine, also asserts that there is no fundamental conflict between theology and science, and it affirms the pursuit of knowledge in all disciplines:

> Though faith is above reason, there can never be any real discrepancy between faith and reason. Since the same God who reveals mysteries and infuses faith has bestowed the light of reason on the human mind, God cannot deny himself, nor can truth ever contradict truth. ... Consequently, methodical research in all branches of knowledge, provided it is carried out in a truly scientific manner and does not override moral laws, can never conflict with the faith, because the things of the world and the things of faith derive from the same God. The humble and persevering investigator of the secrets of nature is being led, as it were, by the hand of God in spite of himself, for it is God, the conserver of all things, who made them what they are.[44]

The 'Crab' Nebula seen from the Vatican Observatory

The Vatican Observatory, Castel Gandolfo, Italy

Sir John Polkinghorne: Theoretical physicist and Anglican priest

"We can take with absolute seriousness all that science can tell us and still believe that there is room left over for our action in the world and for God's action, too," affirms Sir John Polkinghorne, a British scientist, a priest and the author of many books about science and Christian belief.[45]

Professor Polkinghorne worked for many years at Cambridge University, where he was Professor of Mathematical Physics until he resigned to study theology and become an Anglican priest. In his scientific work, he played a role in identifying quarks, the smallest types of particles. He has been honoured with degrees from many universities, he is a Fellow of the Royal Society, and in 2002 he received the prestigious Templeton Prize for science and religion.

Reflecting on the world of quantum mechanics, Professor Polkinghorne says that working in science teaches you how surprising the physical world is. "The quantum world is totally different from the world of every day," he says.

It's cloudy, it's fitful, you don't know where things are, if you know what they're doing. If you know what they're doing, you don't know where they are. So it's a complex world and quite different from what we expected. But it's an exciting world because it turns out we can understand it, and when we do understand it, we have a deep intellectual satisfaction.

There's a deeper, stranger, and more satisfying story to be found, both in science and in religion.

Just as quantum science is not easy to understand, Professor Polkinghorne warns that in Christianity we should not expect God to be simple and easily understood:

High energy particles collision

If the physical world surprises us and is different from everyday expectation – common sense, if you like – it wouldn't be very odd, really, if God also turned out to be rather surprising. Things that are just on the surface, easy to believe, are not the whole story. There's a deeper, stranger, and more satisfying story to be found, both in science and in religion.

Professor Polkinghorne explains that – like God – subatomic particles such as quarks are unseen realities. "Nobody has ever isolated a single quark in the lab," he says. Nevertheless, scientists believe in quarks because they are part of a picture of reality that makes the most sense. "Assuming that they're there makes sense of great swaths of physical experience," he states. "I, in common with all particle physicists, believe very fervently, in a way, in the reality of quarks. But it's an unseen reality. It's the fact that they give intelligibility to the world that makes us believe that they're actually there."

Polkinghorne applies the wave–particle duality of subatomic particles to a Christian understanding of Jesus Christ:

If you're a Christian theologian, and you're telling that sort of story about light being both particle-like and wave-like, we know that the Christian story about Jesus Christ is that he is, of course, a human being but also, in some real sense, needs to be described in terms of divine language. And it's the same sort of dilemma, if you like, and we're not quite so clever,

> **For this scientist and minister of religion there is a consistency between religious understanding of God and scientific knowledge of the processes of nature.**

theologically, at finding the precise answer to that. Again, we don't make progress by denying our experience.

When confronted with the 14-billion-year age of the universe, Professor Polkinghorne talks about the subtlety and patience of God who "works through process and not through magic; not through snapping the divine fingers. And I think that's what we learn from seeing the history of creation as science has revealed it, and I think that tells us something about how God acts generally."

God's patience is an aspect of being a God of love according to Professor Polkinghorne. Love works "not by overwhelming force, but by, if you like, persuasive process. So I think we learn something really quite valuable from that." For this scientist and minister of religion there is a consistency between religious understanding of God and scientific knowledge of the processes of nature: "You can't deduce one from the other, but you can see it and they fit together in a way that makes sense. They don't seem to be at odds with each other."

Golden wave particles

Conclusion

This book has explored some aspects of the relationship between two of the most powerful forces that have shaped our world today, especially in the West. Modern science matured in a Christian context and Christians, for the most part, have affirmed the pursuit of scientific knowledge about the natural world in which we find ourselves.

CONCLUSION

However, Christian belief goes further than science. Christianity places science in a theological context or worldview, affirming that all truth is God's truth and that science reveals just a part of the wonders of the created order, whose ultimate author is God (Psalm 19:1). For thousands of years, the biblical authors have affirmed with the psalmist that "the heavens are telling the glory of God". Many of the giants of the scientific revolution were driven by the same attitude, convinced that God is the ultimate 'author' of both books: the book of Scripture and the book of creation that science interprets.

Unfortunately, rumours of profound conflict began to spread about this relationship between science and Christianity. More than just disputes and misunderstandings, some people have made claims that there are irreconcilable differences between the parties and that we must choose between either a serious commitment to mainstream science or committed Christian belief.

We have seen that an examination of history and also of the nature of science itself reveals that the reports of an inevitable and necessary divorce between science and faith were ill-founded. The relationship is on solid ground as long as we pay heed to St Augustine's caution to Christians over 1500 years ago to take 'science' seriously and to beware of reading Scripture as if it were making 'scientific' statements.

We also saw that science has inevitable limits. On the one hand, science rests on unprovable assumptions and, on the other hand, science is not able to offer answers to the most important questions we can ask. Existential and philosophical and moral questions lie outside the boundaries of valid science. Why are we here? What happens after death? Is there a God? Is there a right and wrong? ... Such questions have no 'scientific' answers and they force us either to look elsewhere other than science or to say, with many atheist naturalists, that there are no answers because such questions are meaningless.

If we agree with atheist philosopher Daniel Dennett that "science is the only game in town", we accept a bleak view of human existence. It means accepting that the human search for purpose and existential significance is fruitless. On this thinking, according to Professor Dawkins, we live in a pitiless universe in which there is, ultimately, "no design, no purpose, no evil and no good". In the opinion of biochemist and Nobel Prize winner Jacques Monod, science has brought the human species to the 'maturity' of accepting this truth about ourselves: "Man knows at last that he is alone in the universe's unfeeling immensity, out of which he emerged only by chance. His destiny is nowhere spelled out, nor his duty."[46]

However, as we have seen in this book, and as philosophers, theologians and scientists have affirmed for centuries, these conclusions do not follow from rigorous science.

Christianity offers a very different but complementary view of humans and their place in the cosmos, while at the same time affirming the vital role of science in revealing the wonders of the natural world. That worldview affirms what science cannot. Humans are not only the pinnacle of evolution on our planet; they are also made in the image of the Creator of this wonderful universe. Humans are crafted for a purpose. That purpose can be summed up as caring relationships: humans are made for caring relationships with one another, for relationship with their Creator, and also for living harmoniously with all creation.

For Christianity, all truth is God's truth; so Christians can rightly embrace the truth that science reveals, as well as holding firmly to the truth that Jesus Christ is Lord of the entire cosmos including this blue planet, home of the only life we know in the universe.

End notes and resources

END NOTES AND RESOURCES

1. St Augustine, *The Literal Meaning of Genesis*, 2 vols, translated and annotated by John Hammond Taylor, SJ, Newman Press, New York, 1982, vol. 1, pp. 42–43.
2. Collins, F 2006, *The Language of God: A Scientist Presents Evidence for Belief*, Free Press, New York, p. 58.
3. Galilei, G 1957, *The Assayer*, abridged and translated by Stillman Drake, p. 4, <https://web.stanford.edu/~jsabol/certainty/readings/Galileo-Assayer.pdf>.
4. Dennett, D 1995, *Darwin's Dangerous Idea: Evolution and the Meanings of Life*, Penguin Books, London, p. 21.
5. F=Gm1m2/r2, or that the power of attraction between two bodies is proportional to the product of their masses divided by the square of the distance between them.
6. Quotations of Graeme Clark are from an interview with Chris Mulherin conducted on 26 April 2017. Excerpts from the interview were published in *Eternity News*, 22 September 2017. See <https://www.eternitynews.com.au/culture/for-the-love-of-science-and-god/>.
7. The source of this common quotation is uncertain. It is quoted in full without a source at <https://www.newworldencyclopedia.org/entry/Johannes_Kepler>. The first sentence is a variation of a phrase found in "Letter no. 117, Johannes Kepler to Johannes Georg Herwart von Hohenburg, April 1599", in *Johannes Kepler Gesammelte Werke*, edited by Max Caspar, Munich, Beck, 1945, vol. 13, p. 309. The second sentence is quoted in Henry Morris, 1998, *Men of Science/Men of God: Great Scientists Who Believed the Bible*, Green Forest, AR, Master Books, pp. 21–22.
8. Lemaître, G 1958, "The Primeval Atom Hypothesis and the Problem of the Clusters of Galaxies," in *La Structure et l'évolution de l'univers Rapports et Discussions. Onzième Conseil de physique tenu à l'Université de Bruxelles du 9 au 13 Juin 1958*, Bruxelles, pp. 1-25. R. Stoops, discussion, pp. 26–31.
9. Darwin, C 1882, *On the Origin of Species*, sixth edition, with additions and corrections to the 1872 edition, John Murray, London.
10. Davies, P 2003, "A Brief History of the Multiverse", *New York Times*, 12 April, <http://www.nytimes.com/2003/04/12/opinion/a-brief-history-of-the-multiverse.html>.
11. Trefil, J 1997, *101 Things You Don't Know About Science and No One Else Does Either*, Mariner Books, New York, p. 15.
12. Dawkins, R 1995, *River Out of Eden: A Darwinian View of Life*, Basic Books, New York, p. 133.
13. Hawking, S and Mlodinow, L 2010, *The Grand Design*, Bantam Books, New York, p. 227.
14. Rees, M 2001, *Our Cosmic Habitat*, Princeton University Press, Princeton NJ, p. xix.
15. Quotations of Jennifer Wiseman are taken from, "Are Religion and Science Always at Odds? Here Are Three Scientists that Don't Think So", *ABC Science*, 24 May 2019, <https://www.abc.net.au/news/science/2018-05-24/three-scientists-talk-about-how-their-faith-fits-with-their-work/9543772>.
16. Mulherin, C 2014, "Genetics – One Way God Speaks to Us", *The Melbourne Anglican*, October.
17. Pope Francis, *Laudato Si': On Care for Our Common Home*, <http://w2.vatican.va/content/francesco/en/encyclicals/documents/papa-francesco_20150524_enciclica-laudato-si.html>.
18. Quotations from Professor Hulme are from a public lecture entitled "Restructuring Climate Policy for a Partisan Era", 5 May 2011, <https://www.youtube.com/watch?v=LqJVHzxWQDU at 3 min 30 sec>.
19. IPCC 2014, *Climate Change 2014: Synthesis Report. Contribution of Working Groups I, II and III to the Fifth Assessment Report of the Intergovernmental Panel on Climate Change*, <https://ar5-syr.ipcc.ch/ipcc/ipcc/resources/pdf/IPCC_SynthesisReport.pdf>.
20. Polanyi, M 1958, *Personal Knowledge: Towards a Post-Critical Philosophy*, Routledge and Kegan Paul, London.
21. Feynman, R 1955, "The Value of Science", public address at the National Academy of Sciences (Autumn) <https://archive.org/stream/feynman_201604/feynman_djvu.txt>.
22. Berry, W 2001, *Life Is a Miracle*, Counterpoint, Washington DC, p. 99.
23. Corney, P 2018, "The Windowless Room", *Essentials*, Autumn, p. 18. See <https://www.efac.org.au/images/docs/EssAutumn18.pdf> and <http://petercorney.com/2017/02/14/scientific-materialism-the-windowless-room/>.
24. Dawkins, R 1995, *River Out of Eden: A Darwinian View of Life*, Basic Books, New York, p. 133.
25. Byrnes, S 2006, "When it Comes to Facts, and Explanations of Facts, Science Is the Only Game in Town" (interview with Daniel Dennett), *New Statesman*, 10 April, <https://www.newstatesman.com/node/164091>.
26. Quotations from Marcelo Gleiser are from: Lee Billings 2009, "Atheism Is Inconsistent with the Scientific Method" (interview with Marcelo Gleiser), *New Scientist*, 20 March, <https://www.scientificamerican.com/article/atheism-is-inconsistent-with-the-scientific-method-prizewinning-physicist-says/>.
27. Collins, F 2006, *The Language of God: A Scientist Presents Evidence for Belief*, Free Press, New York, p. 51.
28. Ruse, M 2007, endorsement for Alister McGrath and Joanna Collicut McGrath, *The Dawkins Delusion*, InterVarsity Press, Downers Grove, IL.
29. Hume, D 1777, *An Enquiry Concerning Human Understanding*, para. 132, <https://www.gutenberg.org/files/9662/9662-h/9662-h.htm>.
30. Mulherin, C 2013, "Interview with Lawrence Krauss", 16 August, <http://iscast.org/kraussinterview>.
31. Draper, JW 1874, *History of the Conflict between Science and Religion*, Appleton, New York, preface.
32. White, AD 1896, *A History of the Warfare of Science with Theology in Christendom*, 2 vols, Appleton, London.
33. Pope Francis, Plenary Session of Pontifical Academy of Sciences, 27 October, 2014, <http://www.academyofsciences.va/content/accademia/en/magisterium/francis/27october2014.html>.
34. Krauss, L 2012, *A Universe from Nothing: Why There Is Something Rather than Nothing*, Free Press, New York.
35. Quotations from Lawrence Krauss are from Chris Mulherin 2013, "Interview with Lawrence Krauss", 16 August, <http://iscast.org/kraussinterview>.
36. Mulherin, C 2012, "The 'God Wars' and the Global Atheist Convention", *Kairos Catholic Journal* 23, no. 5, <https://www.cam.org.au/News-and-Events/Reflections/Article/8252/The-God-wars-and-the-Global-Atheist-Convention>.
37. Collins 2006, p. 200.
38. All quotations by Peter Harrison are from: Chris Mulherin, 2017, "Science and Faith have a Complex Historical Relationship" (interview with Peter Harrison), 16 August, <https://www.iscast.org/interview/Harrison_interview_2017>. See also Peter Harrison's lecture "Clearing the Air about the Galileo Affair" at <http://iscast.org/resources/Harrison_P_2018-03_Clearing_the_air_about_the_Galileo_affair>.
39. Collins 2006, p. 211.
40. Collins 2006, p. 3.
41. Collins 2006, pp. 140–41.
42. Collins 2006, pp. 6, 198, 210.
43. Pope Leo XII 1893, *Providentissimus Deus*, para 18, <https://w2.vatican.va/content/leo-xiii/en/encyclicals/documents/hf_l-xiii_enc_18111893_providentissimus-deus.html>.
44. Catholic Church, Catechism of the Catholic Church, para 153, <http://www.vatican.va/archive/ENG0015/_PX.HTM>.
45. Quotations from John Polkinghorne in this article are taken from Krista Tippert 2005, "Quarks and Creation" (interview with John Polkinghorne), <https://onbeing.org/programs/john-polkinghorne-quarks-and-creation/>.
46. Monod, J 1970, *Chance and Necessity: An Essay on the Natural Philosophy of Modern Biology*, translated by Austryn Wainhouse, Vintage Books, New York, p. 180.

Alexander, D 2003,
Rebuilding the Matrix: Science & Faith in the 21st Century,
Zondervan, Grand Rapids.

Alexander, D 2014,
Creation or Evolution: Do We Have to Choose?
New expanded edition, Monarch, Oxford.

Ames, S, Ashby, R, Mulherin, C & Pilbrow J 2018,
A Reckless God? Currents and Challenges in the Christian Conversation with Science,
Morning Star, Melbourne.

Birkett, KR 1997,
Unnatural Enemies: An Introduction to Science and Christianity, Matthias Media, Sydney.

Birkett, KR 2008,
"I Believe in Nature: An Exploration of Naturalism and the Biblical Worldview", available at
<https://tgc-documents.s3.amazonaws.com/cci/Birkett.pdf>.

Bouma-Prediger, S 2010,
For the Beauty of the Earth: A Christian Vision of Creation Care,
Baker Academic, Ada, MI.

Buxton, G, Mulherin, C & Worthing, M 2018,
God and Science in Classroom and Pulpit,
Wipf & Stock, Eugene.

Campbell, Heidi A & Looy, Heather, 2009,
A Science and Religion Primer,
Baker Academic, Grand Rapids, MI.

Chapell, DF & Cook, ED (eds) 2005,
Not Just Science: Questions Where Christian Faith and Natural Science Intersect,
Zondervan, Grand Rapids, MI.

Collins, F 2007,
The Language of God: A Scientist Presents Evidence for Belief,
Free Press, New York.

Craig, WL & Meister, C (eds) 2010,
God is Great; God is Good: Why Believing in God is Reasonable and Responsible,
IVP, Downers Grove, IL.

Gingerich, O 2006,
God's Universe,
Harvard University Press, Cambridge, MA.

Haarsma, Deborah 2011,
Origins: Christian Perspectives on Creation, Evolution, and Intelligent Design,
FaithAlive Christian Resources,
Grand Rapids, MI.

Harper, C Jr (ed.) 2005,
Spiritual Information: 100 Perspectives on Science and Religion.
Templeton Foundation Press,
West Conshohocken, PA.

Harrison, P 2015,
The Territories of Science and Religion,
University of Chicago Press, Chicago.

Hart DB 2013,
The Experience of God: Being, Consciousness and Bliss,
Yale University Press, New Haven.

Hart, DB 2009,
Atheist Delusions: The Christian Revolution and Its Fashionable Enemies,
Yale University Press, New Haven.

Hitchens, P 2011,
Rage Against God: How Atheism Led Me to Faith,
Bloomsbury, London.

Jaki, S 1978,
The Road of Science and the Ways to God,
University of Chicago Press.

Jeeves, M & Berry, R 1998,
Science, Life, and Christian Belief,
IVP/Apollos, Downers Grove, IL.

Lennox, JC 2011,
Gunning for God: Why the New Atheists Are Missing the Target,
Lion Books, Oxford.

Lewis, CS 2015,
Mere Christianity,
HarperOne, San Francisco.

Lewis, CS 1947,
Miracles,
Macmillan, New York.

Livingstone, DN 2001,
Darwin's Forgotten Defenders: The Encounter Between Evangelical Theology and Evolutionary Thought,
Regent College Publishing, Vancouver.

McGrath, A & McGrath, JC 2011,
The Dawkins Delusion? Atheist Fundamentalism and the Denial of the Divine,
IVP, Downers Grove, IL.

McGrath, AE 2009,
A Fine-tuned Universe: The Quest for God in Science and Theology: The 2009 Gifford Lectures,
Westminster John Knox Press, Louisville.

McGrath, A 2006,
The Twilight of Atheism: The Rise and Fall of Disbelief in the Modern World,
Random House, NY.

McLeish, T 2014,
Faith and Wisdom in Science,
Oxford University Press, Oxford.

Medawar, PB & Smith, B 1984,
The Limits of Science,
Oxford University Press, Oxford.

Numbers, R (ed.) 2010,
Galileo Goes to Jail and Other Myths about Science and Religion,
Harvard University Press, Cambridge, MA.

Plantinga, A 2011,
Where the Conflict Really Lies: Science, Religion and Naturalism,
Oxford University Press, NY.

Polkinghorne, JC 2007,
Exploring Reality: The Intertwining of Science and Religion,
Yale University Press, New Haven.

Polkinghorne, JC 2010,
One World: The Interaction of Science and Theology,
Templeton Foundation Press,
West Conshohocken, PA.

Polkinghorne, JC 2014,
The Faith of a Physicist. Reflections of a Bottom-up Thinker,
Princeton University Press, Princeton, NJ.

Polkinghorne, JC 2007,
Exploring Reality: The Intertwining of Science and Religion,
Yale University Press, New Haven.

Russell, C 1993,
Crosscurrents: Interactions Between Science and Faith,
Regent College Publishing, Vancouver.

Sire, J 2009,
The Universe Next Door: A Worldview Catalogue,
5th revised edition, IVP, Downers Grove, IL.

Spencer, N & White, R 2007,
Christianity, Climate Change, and Sustainable Living,
SPCK, London.

Ward, K 2013,
Pascal's Fire: Scientific Faith and Religious Understanding,
Oneworld, London.

END NOTES AND RESOURCES

There are numerous books, organisations and online resources for exploring the relationship between science and Christianity. Below is a brief list of resources to start the journey.

In Australia, ISCAST – Christians in Science and Technology is run by Christians who are scientists and other academics, with the aim of promoting the science–faith conversation. It has resources and contact details, which can be found on its website at <http://iscast.org>.

In the United Kingdom, the Faraday Institute at Cambridge University at <https://faraday-institute.org> has many resources. Of note are the 20 Faraday Papers at <https://faraday-institute.org/Papers.php>, which provide the general reader with accessible and readable introductions to the relationship between science and religion. Also, the Faraday Institute's Test of Faith video series and discussion guides can be purchased at <http://testoffaith.com> with more resources available online.

In the United States, United Kingdom and New Zealand there are organisations for Christians in the sciences that run events and also have general resources on their websites. See <https://network.asa3.org> and <http://cis.org.uk> and <http://nzcis.org>.

BioLogos, started by Dr Francis Collins (see Article #15, "Dr Francis Collins" on page 58) is a large US organisation dedicated to showing the harmony between mainstream biology and Christian belief. The website, at <http://biologos.org> offers many resources on the topic.

The Center for Theology and the Natural Sciences (CTNS), also in the USA, at <http://ctns.org/> is another significant organisation dedicated to the creative, mutual interaction between theology and the natural sciences. CTNS publishes the journal *Theology and Science*.

The Vatican Observatory at <http://vaticanobservatory.va> is one of the oldest astronomical research institutions in the world, established in its present form by Pope Leo XIII in 1891 "so that everyone might see clearly that the Church and her Pastors are not opposed to true and solid science".

www.ingramcontent.com/pod-product-compliance
Lightning Source LLC
Chambersburg PA
CBHW042239180426
43199CB00040B/2930